CW01261535

THE FORECAST

An Inside Look at How Successful Leaders Drive Revenue Growth

A Business Fable

RICK RALSTON

Visit the author's website at www.RickRalston.com.

Copyright © 2024 by Rick Ralston

Published by Tela Publishing Company LLC, Wausau, Wisconsin

All rights reserved.

No part of this publication may be reproduced, stored in a retrieval system, transmitted in any form or by any means—electronic, mechanical, digital, photocopy, recording, or any other—except brief excerpts for purposes of radio, television, or digitally published review—without prior written permission from the publisher.

No part of this publication may be used in the development or training of any artificial intelligence technology or model—machine learning, deep learning, large language models— including but not limited to applications such as computer vision, natural language processing, generative pre-trained transformers—or any other—without prior written permission from the publisher. The publisher reserves all rights to license uses of this work for development and training of any artificial intelligence technology.

For permission requests and other inquiries, contact the publisher at info@telapublishing.com.

ISBN: 979-8-9875204-1-3 (paperback)
ISBN: 979-8-9875204-2-0 (hardcover)
ISBN: 979-8-9875204-3-7 (ebook)
ISBN: 979-8-9875204-4-4 (audiobook)

All names, characters, businesses, products, buildings, locales, and incidents portrayed in this book are fictitious. No identification with actual persons (living or deceased), entities, places, events or products is intended or should be inferred.

This book is not intended to be a substitute for professional financial advice. The reader should consult their financial professional when implementing the strategies discussed in this book.

Book design and illustrations by Adrienne Blackwell
Author photo by Chelsie Wheeler
Audiobook narration by Jonathan Broscious

First Edition

Tela Publishing
P.O. Box 425
Wausau, WI 54402

www.telapublishing.com

This book is dedicated to my wife, Robin.

For all the years of encouragement and especially for the note on the bathroom mirror.

CONTENTS

	List of Illustrations	ix
	Preface	xi
Chapter 1	THE INVESTOR	3
Chapter 2	THE WINNING WAY	15
Chapter 3	THE JOURNEY	29
Chapter 4	THE FOUNDATION	43
Chapter 5	THE MILESTONES	53
Chapter 6	THE FUNNEL	67
Chapter 7	THE CUSTOMER	79
Chapter 8	THE PURPOSE	95
Chapter 9	THE MISSING PROCESS	119
Chapter 10	THE FINISHING TOUCH	139
Chapter 11	THE FORECAST	147
	Afterword	157
	Appendix	163
	To Learn More	177
	About the Author	179

LIST OF ILLUSTRATIONS

WHITEBOARDS

Revenue Forecast Comparison Chart A	23
Revenue Waterfall Chart	25
Customer Journey Cycle	38
Development Timeline	63
Marketing Funnel	76
Customer Success Process	91
Vision Pyramid	106
Sales Process	135
Vision to Forecast Flow	140
Revenue Forecast Process	145
Forecasting Process	149
Strategies & Metrics	150
Revenue Forecast Comparison Chart B	154

MEETING NOTES

Income Statement & Revenue Charts	27
Forecasting & The Customer Journey	41
Forecasting & Development Milestones	66
Forecasting & The Marketing Funnel	78
Customer Success Plan	94
The Vision Test	110
Sales Process Clarity	138

PREFACE

Building a financial forecast for a business can be a daunting task, even for the most seasoned CEO. There are so many variables and assumptions to sort out. Then there is the inevitable time constraint. After all, the next board meeting is always right around the corner.

If the forecast is too low, the board is skeptical of sandbagging or not being aggressive enough. If the forecast is too high, there is the increased risk of missing targets. A couple of missed quarters and the pressure is on. Fix it or you're out!

Sure, there's the fast and easy route of taking last year's results and adding a nice round percentage. That method is hard to support, though, once the questions start flying in like crop dusters on a Midwestern cornfield.

Or how about the top-down method of management dictating the targets without team input. Did that ever work? All CEOs know this method doesn't work and never has. But due to the lack of a better method and the lack of time, it still happens, and unfortunately, quite often.

Then there's the "just add more salespeople" approach. Yeah, right. If that worked, every business would be wildly successful and there would never again be an unemployed salesperson.

Over the years, I've seen a lot of different methods for building forecasts. Some have been effective. Most have not. One thing I know from sitting on both sides of the table in the boardroom is that it is a lot easier for a board member to shoot holes in a forecast than it is for a CEO to build and present an effective forecast.

This book presents a different forecasting process. One based on logic and built by a team of people holding themselves and each other accountable. The process described takes time and dedication. It is not easy—or fast. However, when orchestrated with the right leadership skills, much more than just an effective forecast will be produced. A stronger, more resilient team will emerge, a team that can use their forecast as a tool with which to learn and improve.

True growth is more than forecasting numbers.

I hope you find this story enjoyable, but even more, educational.

THE FORECAST

CHAPTER 1

The Investor

"Hey, Ted," Danny said as he hit the top of the long wooden staircase in Ted's office.

Ted hesitated at first, because he was not alone. "Uh, hi, Danny. Uh, this is Mr. Larobie, the investor I mentioned. Mr. Larobie, this is my friend Danny." Ted's voice was nervous. Danny had interrupted the wrap-up to an important meeting.

"Good to meet you, Danny," replied Mr. Larobie in a confident voice. His handshake was strong like a farmer's. Speech polished. It came out like he wasn't even trying. He had the elusive executive presence business books talk about. "Please meet my associate Phillip too. We were just heading out. Thanks again for your time today, Ted. We're looking forward to hearing about your next release."

Ted shook his head as Mr. Larobie and Phillip headed down the long stairway. The introduction was short. Danny was early to meet Ted for lunch, and the quarterly progress meeting had run over.

Mr. Larobie looked to be in his eighties. You could tell from the way he carried himself and his dress that he was confident. Someone with experience probably learned the hard way. He had a certain smile and spark in his eye that was comforting, like he knew the answer before you even asked the question. Phillip was much younger—mid- to late thirties, most likely—still older than the early twenties of Ted and Danny, though.

Ted's building was old brick with hardwood floors. There was always a smoky stench of open grill wafting up through the floor from the Wasabi Palace Japanese restaurant downstairs. But what stood out to Danny most about Ted's office was the click of his sales manager's shoes. Her name was Jessie, and she was high energy, all the time. *Click, click, click* down the hall. Always a huge smile and a firm handshake. That alone could close deals, Danny thought. He knew his family business could use some of that energy in its office. But what did he know? It was only his second day on the job. Any job, really.

Off to lunch.

Downstairs in the Japanese place to grab some sushi, Ted had to correct Danny and point out the difference between sushi, sashimi, and all the other assorted names. He kind of came across as arrogant at times, but Danny was used to it. They had been buds since grade school.

There was a second room off to the side. It was darker and less noisy than up front at the sushi counter. They always sat at the same table when Danny visited Ted's office. Usually talked about football or baseball, whatever the season was.

Today, though, Ted was tense. "How was your meeting?" Danny asked.

"Not good," Ted replied without any hesitation. It was like he didn't want to talk about it, then he immediately started talking about it. Ted was not usually this open about his business.

He had bought the technology a couple of years back, after dropping out of college. He had been a business major but left early to try his hand at entrepreneurship by starting up a financial technology business. He called it fintech, and he had a good pitch. Ted could usually convince anybody to do anything.

Ted was tall and athletic. He played college baseball until he hurt his shoulder. He never said so, but rumors were that losing his scholarship was what made him decide to leave school. Ted frequently told Danny all real entrepreneurs quit school early anyway.

Ted had convinced Mr. Larobie to invest some money so he could buy a check-scanning software company out of Florida. The technology was old, but it had lots of existing customers using the software. Ted thought he could get the code rewritten and updated, then sell the business and make a mint. Danny's dad said it would never work. He said the tech was all antiquated and would be worthless soon if Ted didn't keep up the innovation. Danny trusted his dad. After all, his dad had been building software Danny's whole life.

Ted started to vent. "Mr. Larobie doesn't know my business. He doesn't know anything about software. It's an embarrassment, the questions he asks. He's a real estate guy, not a software guy!"

Wow, Ted was really upset. Danny had to ask him to slow down and quiet down too—they were in public, after all, and everyone here knew Ted, and probably Mr. Larobie too.

"Calm down, buddy. What's going on here?" Danny asked.

"He comes in with his dang yellow folder full of notes and questions. If he would just listen to me on what we need to do and put some more money into the business, everything would be fine. The next software release is going to be awesome!"

Ted was in a bind. Not as serious as the bind Danny was in, though. Danny and Ted were meeting today to talk about Danny's family business. Not Ted's business. A few months back, Danny's dad had passed suddenly, and his sister had been trying to keep the wheels on the business. It wasn't going well.

Sis was a full eight years older than Danny. Growing up, their folks used her as a built-in babysitter for Danny. She had been doing the sales for years for the company. Danny assumed she didn't know much more about the rest of the business, though. They were in a financial crisis. No growth and no cash.

Danny's dad had built a software company called Red Line. It was software that counted the people who got on and off public buses and trains. That way, the city transportation groups could forecast stops and traffic to be more efficient. He and Danny's uncle Bill had invented and patented a special laser technology. Uncle Bill left to go to work in a much larger software company. Danny always thought Uncle Bill didn't like working in a small company with all the financial problems, like the ones they were having right now.

Danny thought Ted could give him some advice. It was just Danny's second day working with Sis, and he was honestly lost. His last semester at college had just ended. He had skipped graduation to come help her. It wouldn't have been a good graduation anyway, without his dad. Sis admitted she was in over her head in the business and didn't know where to turn next. Danny thought maybe she was just saying that to get him to come home. She was not normally that nice to him.

After Ted calmed down, Danny decided not to bother him with his problems today. Instead, Danny asked if he would mind giving him Mr. Larobie's number, thinking Mr. Larobie might know someone who could help. Danny thought surely Mr. Larobie would be connected.

Ted said, "Yeah, sure, but don't expect to get in to see him. He runs a tight ship. That driver of his, Phillip, keeps close tabs on who goes in and out. It might take weeks to get in to see him, if at all."

"Ted, I don't think that's his driver. While you were rambling on a while ago, I looked him up online. Phillip has an MBA from Stanford. He spent several years at an investment bank in Silicon Valley too."

"Yeah, whatever. He drives old man Larobie everywhere and carries that dang yellow folder."

THE INVESTOR

***

It only took a couple days for Danny to get in to see Mr. Larobie. He thought maybe Ted didn't ask nicely enough.

While sitting across from Mr. Larobie's desk, Danny noticed all the business books on his shelves. Based on the rough guess of the books per shelf and the number of shelves, he would estimate over five hundred books. One shelf made Danny curious. It contained about a dozen copies of the same thin black book with no title.

Mr. Larobie's large desk was old wood with a glass top. The desktop was nearly empty, with only a few things on it at all. A family picture. A small clay pot like a child would make at school. Maybe his grandchild. A couple of nice ink pens and two stacks of file folders. One stack was yellow, and the other, taller stack was blue. That was it.

Danny was nervous. Mr. Larobie was wealthy. That was new to Danny. His family had not been rich; they did all right. No country clubs or fancy cars, but they had enough growing up. The family business was small. Only a dozen or so employees. Not huge like the tech companies Ted talked about all the time.

When Mr. Larobie came back in and sat down, Danny started talking like he just couldn't shut up. Maybe it was the coffee on the drive over. Danny asked him about all his books on his shelves and which ones were worth reading. Mr. Larobie admitted he hadn't read them all, at least not cover to cover. He said some he skimmed and tabbed the good points. Danny even asked Mr. Larobie if he had ever written a book.

"I've tried a couple times. Doesn't seem writing suits me as much as meeting and talking with people." His voice was strong and polished. "When I meet with people, I can hear their story and then do a better job connecting them with the best resource to help them be successful."

"What about those little black books over there? What are those?" Danny asked, realizing he had overstepped, and it was probably none of his business.

"Those are books I give out to people I work with."

"What kind of book is it?" Danny really could not stop.

"Let's talk about why you asked to meet today, then maybe I'll tell you about those books. Sound fair?" he said, getting the conversation back on track.

"Sure, that's a good idea." Danny could tell he had asked too many questions already.

They then had a very candid conversation about the dilemma at Red Line. Danny told him all about how, growing up, he wanted to be in the family business, but his dad made him work outside the company. In high school, he painted fenceposts on one of the neighbors' ranches. During college he tried his hand at starting his own lawn-mowing business. Danny loved the work, and he loved talking to people, and he loved learning about business even more. Unfortunately, he never learned much from his dad about the family business. Danny thought maybe his dad must have wanted something better for him. Or at least different.

"After our dad passed, our family accountant told my sister we should consider selling the business or raising money to keep it from going under. She admitted to me there had not been enough sales, and she blamed herself for that. Selling the business didn't feel like it was the right answer for us. Or at least it shouldn't be the only option. Sis asked me to join her in the business to help her get it back on track, and I agreed. We thought a second opinion might be the right next step. That's why I'm here."

"So, you need cash to keep the business running?"

"Well, that's the thing—we don't know. We do know we need to learn more about finances, and I remember hearing Dad say throwing cash at things doesn't always fix them."

"Your dad was a smart man."

"You knew my dad?"

"I met him once or twice, years ago."

That seemed a little odd to Danny, since his dad was never into real estate, only tech stuff, mainly hardware and software. He blew it off, thinking it was a small world kind of thing.

Once it was all on the table, Mr. Larobie agreed to help Danny, on one condition.

"Danny, I like how you are committed to learning instead of just jumping to an answer. I will agree to look over your business with you later next week if you have a forecast for Phillip and me to review. Do you have a forecast?"

Oh no, Danny thought. He didn't think they had one. But how tough could it be? He just graduated with a degree in physics. Building a forecast should be a piece of cake compared to that.

"Sure, no problem, that would be great. We will have a forecast for you next week!" Danny said. "Oh, I almost forgot, what about those little black books?"

"Oh, yes, it is like a reference book to help people with business and financial challenges. I call it *By the Numbers*."

"Is that what that book is? A book to teach business and finance?"

Mr. Larobie replied, "No, but that would be a good book too, wouldn't it? This is simply a book of numbers to help you on your learning journey."

Mr. Larobie pulled a black book from the shelf. "You see, it works like this. When you are stumped on a topic, you look it up in the book. Next to each word, instead of a definition, there is a phone number of someone who has already faced and conquered that problem." He opened the book. "All you have to do is text the number and use this simple phrase: 'I have a question about the numbers—will you help me?' For instance, here under the letter R is the word *revenue*." Mr. Larobie pointed to the word revenue in the book. "If you had a question related to revenues, you just text this number right here, and you'll get help."

"Wow, that sounds super easy," Danny said.

Mr. Larobie continued. "Anyone listed in this book will always be there to help you. That's a requirement to be in the book. The book only includes people who are open and willing to learn and help others learn too. It's a fundamental requirement in any business. Learn and help others learn too."

Danny thought if he could just get his hands on one of those books, maybe he and Sis could figure things out from there.

Feeling confident as he stood up, he asked, "Any chance I could get a copy of your *By the Numbers* book?"

"Here you go, Danny. Take this copy. All you had to do was ask, and I hoped you would. One thing, though. The book is for your eyes only. It's between you and me. You can't share it with anyone else. The people in the book dedicate a lot of time and effort, and we must treat their time with the utmost respect."

Danny figured Ted probably already had one of these books, but he knew he should double-check. "Talking to my friend Ted would be okay, though, right, Mr. Larobie?"

"Well, not yet Danny. Let's keep it between us. Ted is quite busy with his own business right now."

That seemed odd to Danny. Maybe Mr. Larobie had already figured out that Ted still thought he had all the answers. Ted was not at all like what Mr. Larobie had said earlier about people being open and willing to learn and help others.

"One more thing, Danny. Please call me Sal. Everyone I work with does."

CHAPTER 2

The Winning Way

As Danny left Mr. Larobie's office, he couldn't wait to tell Sis all about the meeting. He knew she would be so excited to have help. Something told him everything was going to be just fine. He knew she would like the part about the *By the Numbers* book too.

Oh, wait. Sal had said it was just between them. Surely, he could tell Sis, though. She was in this deeper than he was.

Danny kept thinking. Nope, the book could wait. He may never use it anyway. Who knows, it might not even work. He decided to just keep it to himself for now.

It was Monday morning of the next week. Danny and Sis had spent the last few days building out what they thought was a good format for a forecast. They had to look it up online. They had worked through the list of existing customers, how much each paid per month, how many would grow, how many would shrink. Sis knew of some customers who were talking to other vendors. They took a stab in the dark at how many new customers she and the rest of the sales team could land. Against her better judgment, Danny convinced her to push up the amount each person had to sell. *How else do you do it?* he thought. Their guess would be as good as anyone else's. She didn't like that idea but went with it anyway.

Danny thought he had something to share for Thursday with Mr. Larobie. It felt off, though. It was still early in the week. He had a couple more days. While sitting in the Red Line conference room, his temporary office, he realized just how old and worn-out things were. They barely even had a whiteboard—if you could call it that. It was stained with old markings from past meetings. The markers were always dried out too. The conference table chairs didn't match. It was a mess.

Danny glanced over and saw the little black book on the table. He picked it up and looked up the term *forecast*, thinking maybe someone could help him. He thought all he needed was someone to look over his shoulder and make sure he was using the right format and didn't look stupid when he gave it to Mr. Larobie.

Danny texted the number next to the word forecast. Just as Mr. Larobie had instructed Danny– "I have a question about the numbers. Will you help me?"–he did it. Sent.

Ping. His phone went off. *No way! That was fast.*

The reply text read, "Hi, this is Julie Winn with Winning Robotics. How can I help you?"

Danny was excited! He loved Winning Robotics. They made cool drones and field crawlers for making outdoor videos. Danny's neighbor with the ranch he worked on as a kid used them to video record his whole place during the flood a couple of years back.

His mind came back to the present as he realized he had to have this discussion now. "Hi, Julie. I'm Danny with Red Line Technologies." He was still getting used to saying the company name like that.

"Hi, Danny. Good to meet you. What do you want to talk about today? I'm here to help any way I can."

"Could you take a look at our forecast?"

"Sure. Want to come by our office in Canyon Creek this afternoon at two?"

Danny was more excited than ever. He thought for more than a minute that the book was indeed magical. He realized why Mr. Larobie wanted to keep it quiet. "Yes, that would be great!" Danny replied.

He decided to take Sis with him, since she had done most of the work on the forecast so far. She was a little reluctant and wondered how he had made the intro to Winning Robotics so fast. He told her something about a new local networking app he was trying out.

***

When Danny and Sis showed up at two in Canyon Creek, Julie seemed eager to help and was happy Danny brought Sis along.

Her real name was Cecily—everyone just called her Sis for short. She was a little timid at times. Danny wondered if that hurt her in her sales role for Red Line. Sis was not like Ted's sales manager, Jessie, the one who clicked her heels down the hallway. Jessie was super outgoing. Always smiling. Sis was just the opposite. Sis was all business. Brass tacks. Matter of fact. Today she introduced herself as Cecily. That was new to Danny. It was the first time he had heard her real name out loud—except when their dad had been mad at her.

Julie gave them the quick tour. Danny wished it had been longer. He was caught up in all the cool stuff being built. It was much cooler than what he thought about Red Line. The whole place was like one big robotics workshop. Julie said something about how they took the robots to local elementary schools to help teach kids about technology. *Lucky kids*, Danny thought.

It was obvious Julie liked her business. She was a techie and a worker. She wore trail runners and blue jeans. You could still see a little red dirt on her running shoes. She could have easily fit in on Danny's neighbor's ranch land, running her robots and drones. He thought maybe she had even been there.

Julie took Danny and Sis into her office, which was no different from the rest of the place. Soldering irons and circuit boards lying around. Wheels and propellers too. Smelled kind of like motor oil. It reminded Danny of the physics lab back in college. He loved it.

"Show me what you've got!" Julie jumped right in.

Sis handed over the forecast. It fit on just a couple of pieces of paper.

Right off the bat, Julie scratched through the word P&L Sis had put at the top. "Why did you scratch that, Julie?" Sis asked. "That's what our accounting software calls this layout."

"Some people call it a P&L. Most software does too. It's also called an income statement. There are minor differences, but the two are mainly the same. The words *profit* and *loss*, for P&L, imply there are, or will be, losses. That's not a healthy way to look at your business. It happens, especially when you take on investment capital, but we need to focus our energy on income, not losses. Sal is picky about this. He wants your energy in positive things, not negative ones."

Sis looked at Danny, clearly wondering who Sal was. Danny knew that she realized he had been up to his little brother tricks again.

"Plus, some companies get used to running in the red, to the point where they forget how to grow. You can't achieve your vision in business when you stay in the red." Julie paused, then added, "Okay, let's see your underlying assumptions for these new customer revenues."

Danny was a little embarrassed, and he wasn't even sure he knew what Julie meant. "There's not a lot to support the new customers just yet. We have a detailed table of data on all the existing customers, though, which is where the money is today. Isn't that what's most important?"

"It's good that you have that, Danny, but the revenue forecast is a combination of both existing and new customers. Plus, it must include expansions and contractions."

"Okay, we have some of that," Sis added. "We have upsells—I guess that is expansions. And what did you call it? Contractions? We estimated who would grow and who would shrink. Is that what you mean?"

"Yes, exactly. Let's also include the assumptions for the new customers too. When a business is growing like I know Sal believes yours will, everything is about new customer growth and existing customer retention. Sal never gives *By the Numbers* to companies and people he doesn't believe in. You're a lucky family."

Danny bit his tongue when Julie mentioned the book. He could tell from Sis's expression she was getting irritated.

Sis asked, "Yeah, I don't know. We don't get many new customers. That is the real problem. No growth."

"You're in good hands with Sal. Keep your chin up. Let's talk more about your forecast," Julie said.

"Okay, Julie, thanks," Sis said. "Everything we found online says this is a good format to use. That's why we tried it. What makes an income statement a good format for forecasting?"

"Good question. Once your forecast is approved by your board of directors as a budget, you will need to show a comparison of budget to actuals each month or quarter. Since your actual results will be in the income statement, that makes this a convenient format. It also works well with accounting software, so the comparison becomes almost automatic. It is also good to show a comparison to the prior year. Want me to whiteboard it for you?"

"Yeah, that would be great, thanks!" Sis said.

Julie said, "Here you go. This is the endgame of a forecast. In this one chart you can see a quarter's worth of top-line revenue broken down by month. It compares the prior year, the forecast, and also has the actual results for each month. In this last period, you can see both the growth year-over-year for the same period, and if you hit or missed your forecast. Hopefully you beat your forecast, like I show here on the right."

"That makes it easy," Danny said. "See, Sis, this isn't so bad."

Sis rolled her eyes at Danny again.

Julie added, "Just because this chart is easy to understand doesn't mean it is easy to create. There are a lot of drivers to get to these numbers. Like all the expansions and

REVENUE FORECAST COMPARISON

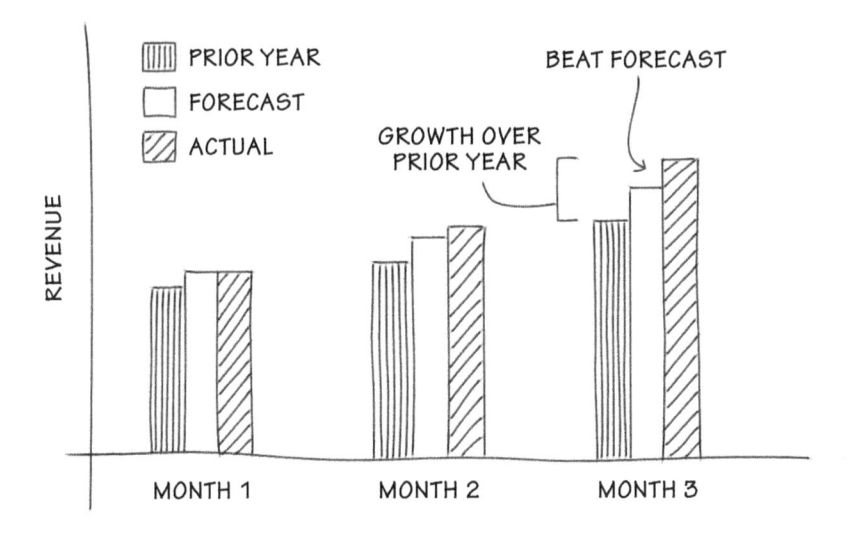

Chart showing revenue comparison across Month 1, Month 2, and Month 3 with Prior Year, Forecast, and Actual bars. Labels indicate "Growth Over Prior Year" and "Beat Forecast."

contractions we were just talking about. Let's go back to that for a second."

Danny looked over and saw Sis was taking notes like crazy. He had never seen her this engaged in anything before. He was starting to see this wasn't just another assignment at school. This was real life.

Julie jumped back up to the whiteboard and drew a new blank chart. "We'll start with the number you have for existing customers. Let's call that *beginning revenue*. That is the revenue you have at the beginning of a period, like a month or a quarter, before anything changes."

"Okay, we've got that number, for sure," Danny said. "We also have that broken down by each customer's name."

"Perfect," Julie continued. "That will be the first column on a revenue waterfall, or some people call it a revenue bridge. The second column is the additional revenue from the new customers you are going to add during the period. We'll just put a blank in there for now. What do you think the next column will be?"

"Uh, maybe the expansions?" Danny guessed.

"Yep, let's add that next," Julie said. "The order doesn't really matter. Just like the new customers, the expansions make the waterfall, or bridge, go up. Some companies really grow a lot just through their expansions. Then comes the contractions, which will bring it back down some. Hopefully not too far down, though."

"Okay, cool. So that gets us to the ending revenue, right?" Danny asked.

Sis jumped in. "Not quite." She paused. You could see in her eyes this was all getting too real. "We lose customers too, Danny. We will have to take out our customers who terminate and go with another vendor."

Julie added, "Yes, sorry to say, Cecily, you are right. The next column is for the revenue you lose from customers

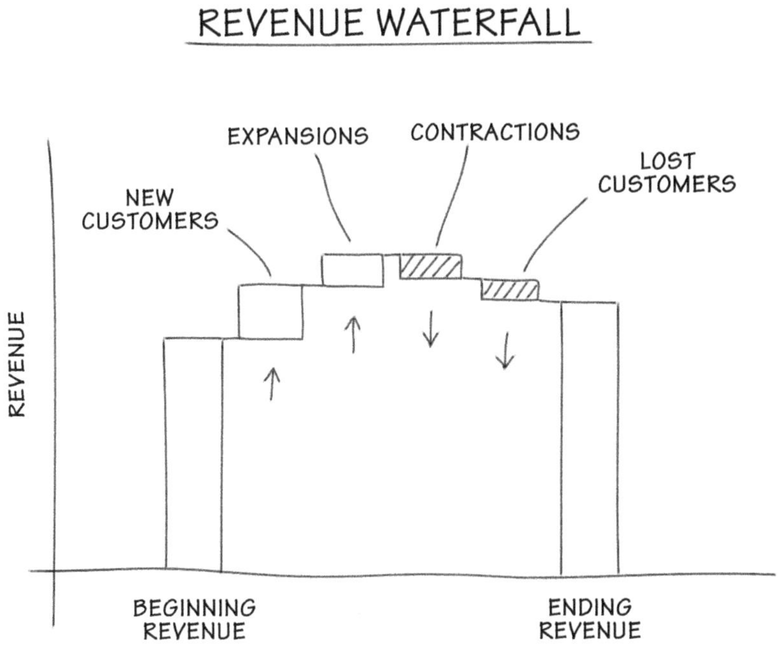
REVENUE WATERFALL
EXPANSIONS
CONTRACTIONS
LOST CUSTOMERS
NEW CUSTOMERS
REVENUE
BEGINNING REVENUE
ENDING REVENUE

will bring the chart back down a little too. The good news is, with this waterfall or bridge approach, you can see which levers you can pull to kick-start the growth."

"Guess there is a little more we are going to have to add, huh, Sis? But we can do this," Danny said, knowing he still needed a lot of help to be ready for his meeting with Sal.

Julie added, "It takes all these levers to build a good forecast. I like to use a waterfall to show how one month rolls into the next, which then rolls into the revenue side of the income statement forecast. Here at Winning Robotics, we like to call our business planning method 'The Winning Way.'"

Sis said, "Julie, thank you so much for taking the time to meet with us today. It has been extremely beneficial."

Danny saw a different side of Sis that day. She came across so polished and professional.

On the way out the door, Julie said, "Hey, Cecily, let's set up a call soon. I would like to talk to you about an upcoming leadership conference here in town."

"Uh, yeah... Sure, let's do that." Sis seemed surprised. Danny knew she hadn't thought of herself as a leader.

Sis slugged Danny in the parking lot and asked, "Hey, bro! What is that stuff about the book *By the Numbers*? And who is Sal?"

JULIE WINN + SIS
WINNING ROBOTICS

MEETING NOTES

"INCOME STATEMENT", NOT P&L
FOCUS ENERGY ON INCOME, NOT LOSSES.

☆ GET BOARD APPROVAL OF THE FORECAST

REVENUE FORECAST COMPARISON:
COMPARE TO ACTUAL INCOME
COMPARE TO PREVIOUS YEAR INCOME

→ ACCOUNTING SOFTWARE MAKES IT FAST + EASY

REVENUE FORECAST CHART

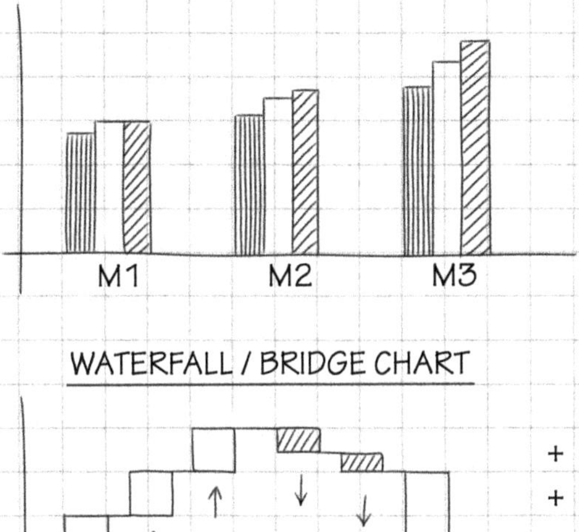

PRIOR VS.
FORECAST VS.
ACTUAL

WATERFALL / BRIDGE CHART

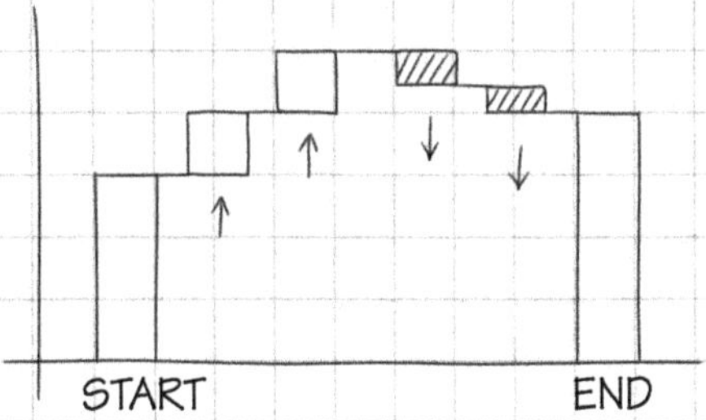

+ NEW SALES
+ EXPANSIONS
- CONTRACTIONS
- LOSSES
= GROWTH

YOU CAN'T "WIN" IF YOU'RE IN THE RED.

CHAPTER 3

The Journey

Thursday's meeting with Sal and Phillip came way too soon for Danny. He wasn't ready. Half of what he learned this week from Julie wasn't in the forecast yet. He was going to have to wing it.

Unsure what would happen on Thursday, Danny reached out to Sal's right-hand man, Phillip, aka "The Driver." Danny asked if it would be all right if he invited Sis and his uncle Bill, who knew the technology and knew how his dad thought. There was no problem, and Uncle Bill agreed too.

The meeting was in Sal's boardroom. The big black leather chairs were old and worn, the tall kind with brass rivets around the edge. The table was long. Danny couldn't quite place it, but there was something that reminded him of the old library at school. Maybe the smell of old textbooks, the big, thick books that are a hundred years old. It made Danny feel important and intimidated at the same time.

The part of the forecast Sis had worked on with Julie went great, even with missing data. All the sales and assumptions about new customers and existing customers worked out better than he planned. Danny thought everyone was impressed and pleased with the forecasted sales. Sis even added some sales from new products they were working on. Danny was grateful to have Sal's book and to have connected with Julie.

Sis had warned Danny before the meeting that delivering on their forecasted numbers was going to be a lot harder than presenting them in a meeting. She knew how hard it was to sell the technology when the competition was rolling out real-time imaging using AI as an alternative to laser counting. She had told Danny all about how Red Line was behind the times.

During a break in the meeting, Danny noticed Uncle Bill and Sal chatting it up in the corner. Maybe they had something in common Danny didn't know about. Maybe Uncle Bill wanted to get into real estate.

Shortly before they all regrouped at the table, Danny walked behind Uncle Bill's chair to get back to his seat. He saw his uncle Bill had a lot of notes on the forecast handout. Danny stopped and asked what was going on. Bill said he had several questions about the R&D expenses, new products, and something about milestones and project timelines.

"Can you hold that 'til another day? I want the meeting to go well. It's our first one with Sal," Danny pleaded.

Uncle Bill nodded.

Honestly, Sis and Danny had not even talked about the new products she added. It was news to Danny. They hadn't talked to Julie about them either.

When they sat back down, the meeting picked up where it had left off, wrapping up the revenue section.

Sal opened up his blue folder and looked across the table at Danny. "How do you feel about your new product R&D expenses and new product delivery dates, Danny?"

After the brief conversation with Uncle Bill just a minute ago, Danny thought hard about how to answer. "We are still working through a couple things to get it right," he said, thinking that would give him some wiggle room.

Uncle Bill looked satisfied Sal had asked so he wouldn't have to.

Sal smiled.

Danny and Sis made it through the rest of the forecast with several ups and downs. Mainly ups. But Danny could tell he still had a lot to learn about financial forecasting, and about business overall. It was a lot different than the physics formulas he had been learning in college.

Afterward, Danny thanked Sal again for the book and told him how he and Sis had met with Julie. Sal appeared pleased Danny included Sis in the meeting and didn't mind that she found out about the book. She was on the same team, after all. It seemed to Danny that Sal already knew they had met with Julie.

Sal walked Danny out and said, "Let's talk again in a week, next Thursday. Just you and me. I've got some ideas for you."

When Danny got home that night, he looked up *R&D expenses* in the *By the Numbers* book, stirred up the nerve, and started to key in the phone number to text. That's when it hit him—the number belonged to his uncle Bill.

***

Tuesday afternoon the week after the first forecast meeting, Uncle Bill was excited for Danny to come visit his office. Danny had been up there once or twice when he was a kid but forgot how big the company was. There were many more people than Danny remembered, and a lot of activity. Everyone was busy, working, smiling, collaborating. Still very corporate, though. A different culture than little Red Line Technologies. That was for sure. Or even Winning Robotics.

Uncle Bill walked Danny into an area he referred to as the R&D team. There were all types of devices set up. Mobile phones, tablets, TVs, PCs, Macs—you name it, they had it.

"We must test our digital photography software on all these devices before it is released to the public. This is also where we come up with new functions and features. For instance, that team over there in the conference room is working on our new AI image-recognition software."

There had to be at least forty or fifty people in this area. Uncle Bill talked about sprints and burn-downs, then pointed to a board across the room with lots of sticky notes on it. Danny could tell Uncle Bill liked being here with this large R&D team.

Uncle Bill asked how Tommy T. was doing. "He was always great to work with. I remember back in the day at Red Line, he could have built our entire product all by himself!" Tommy T. was one of the original engineers who designed the prototypes along with Danny's dad and Uncle Bill. "Tommy T. has a sky-high opportunity cost. You and Sis need to keep him on your most important projects. He's sharp!" Uncle Bill said.

Danny wondered exactly what his uncle Bill was trying to tell him about Tommy T. No way would Uncle Bill steal one of the Red Line employees! Danny knew his dad hated that idea. He would never do that to someone else. And what did Uncle Bill mean by opportunity costs? Danny hoped he could look it up in the little black *By the Numbers* book.

"Danny let's go into the breakout room over here, and we can talk about R&D expenses and project timelines. Sound good?"

"Yeah, let's do it!"

"The R&D team is critical to the timing of new products. That's why Tommy T. is so important to your forecast. If the new product is late, then the sales will be late too. You must be careful, though; new product development teams can get expensive."

Maybe that was where the cash problem was at Red Line, Danny thought.

"Let me show you how we look at the revenue forecasting process here. It takes a lot of input from a lot of different departments," Uncle Bill said. "This is important for the Red Line forecast too."

"Sure, that would be great. Julie over at Winning Robotics already told us how they use a revenue waterfall. It made good sense," Danny said.

His uncle Bill agreed. "Yes, she has it right. The waterfall approach is a good one. We use that too, for our recurring revenue streams."

Uncle Bill went up to the whiteboard in the breakout room and drew a circle and broke it into four quadrants with an arrow going around each one.

"It's what we call the customer journey," Uncle Bill continued. "It starts with what we've just been talking about, and that's the new product development process. You can't market or sell a new product until you have it built. We put this in the upper-right corner."

"Okay, but new products are just a small part of our forecast," Danny said.

"Yes, right now. But those new product sales could be the future of the business. Maybe not in the next quarter or two. But someday. So, anyway, it starts with R&D."

"Got it. What's in the next quadrant?"

"That's marketing. Down in the bottom right. I noticed in the presentation last week there wasn't much talk about marketing to find new prospects. That's a gap. Phillip will point that out to Sal. Phillip pays very close attention to details like that. He is a numbers guy."

"By the way, Uncle Bill, how do you know Sal? Isn't he a real estate guy?"

Uncle Bill almost laughed. "Oh, I thought you knew. Yes, Sal certainly owns a lot of real estate. Including the building we are in right now. He also is involved in a lot of tech businesses, not only here, but across the States."

He continued, "Danny, do you remember when we had a rough go of it a few years ago here in our photography business? Sal stepped in as chairman of our board for a while until we turned it around and we were growing again. He's the best. I'm so glad he is helping you and Sis."

"How does Sal know Dad? He mentioned he had met Dad once or twice."

"Back when your dad, Tommy T., and I started Red Line, before we even called it Red Line, your dad pitched Sal for an investment. Sal made us a good offer too. But in the end, your dad said throwing money—"

"—throwing money at things doesn't always solve the problem."

"That's right. Turns out your dad made a good decision. We landed our first customer shortly after Sal's offer, and the company took off. The business has made a good living for the family for a long time," Uncle Bill said.

Uncle Bill was getting a little emotional. Talking about Dad again was hard for him. Danny too. Real hard. Danny wanted to ask Uncle Bill why he left Red Line. He decided to save that conversation for another day.

Uncle Bill continued, "Back to marketing. Some companies try to forecast their new customer sales by estimating how many new customers or total new dollars each salesperson can bring in. In reality, it doesn't work that way. If it did, we would all just hire more salespeople. A company needs a pipeline of new marketing leads. Once R&D sets a timeline for a new product in the upper-right quadrant, the marketing team is out there building up the list of potential customers for it down here in the lower-right quadrant. Have you and Sis been talking about marketing yet?"

"No, not much. Well, not at all, really. That sounds like it would cost more money, and she says we don't have any. She thinks that's why we don't have growth anymore. To be honest, the lack of growth has turned into a lack of cash. Seems like a vicious circle, chicken and the egg, Catch-22, something or other. That's why we went to Sal in the first place. Not for cash, just for advice. He was helping my friend Ted, so I thought maybe he could help us too."

Uncle Bill said, "I'm glad he is helping you; he and his connections have helped us over the years many times. Let's keep moving. Next in the customer journey is—"

Danny cut in, "Sales! That one is kinda obvious, Uncle Bill."

CUSTOMER JOURNEY

```
CUSTOMER SUCCESS          RESEARCH &
     TEAM                 DEVELOPMENT
                             TEAM

              CUSTOMER

   NEW SALES              MARKETING
     TEAM                    TEAM
```

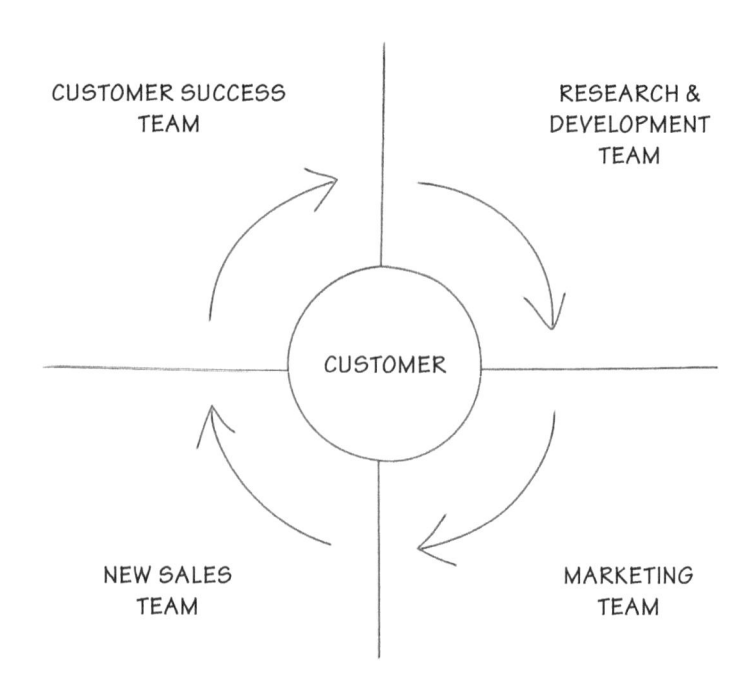

"Yes, it all flows together once you see it. After marketing, sales is down in the bottom left. That's where we close deals. Marketing finds new prospects. Sales closes the deals and gets the contracts signed. It takes a solid sales process with lots of training and discipline for this quadrant to work."

"Last week, most of our forecast was on sales, new customers, and existing customer expansions and contractions, the stuff that Julie helped us with."

"Good lead-in, Danny," Uncle Bill said. "The sales quadrant in our customer journey is for new customers. Existing customers go in the quadrant above it in the upper left. That completes the customer journey. We call it customer success. They take care of keeping the customer happy, renewing contracts, and growing the existing customers. They are also on the hook for customers who shrink or choose to leave."

"So, that's a separate team for your company?"

"Yes, it is here, but not in all companies. We like to keep them separate and focused. That is harder to do when you are a small company. That's it, Danny. The whole customer journey. Each team has their part to contribute to our forecasting process. Then, after the forecast is approved, each of them is responsible for their part too."

"Hate to cut it short today, Danny, but I have a meeting with our R&D team on some of our own project timelines," Uncle Bill said.

"No problem. This has been a huge help. I didn't realize how much there was to forecasting. Sis and I still have a lot of work to do."

"Try to remember the main point today, Danny. It takes a lot of input from a lot of people. Good forecasting takes a team," Uncle Bill said as they walked out together.

Danny was starting to understand why Uncle Bill went to the bigger company now. He figured out his uncle knew a lot about business, not just engineering. Danny also realized he and Sis had a lot to learn. He accepted that they looked naive in the meeting the week before.

"Oh, Danny, congratulations on your graduation. Your dad would be proud. Glad you're back home. He would like that too," Uncle Bill said.

UNCLE BILL
UNCLE BILL'S OFFICE

MEETING NOTES

FORECASTING TAKES LOTS OF INPUTS
FROM LOTS OF PEOPLE

☆ THE CUSTOMER IS ON A JOURNEY
IT'S OUR JOB TO HELP THEM ALONG THE WAY

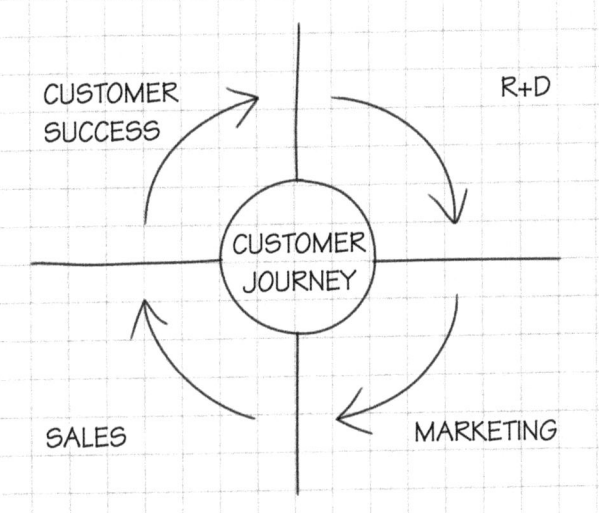

```
        CUSTOMER              R+D
        SUCCESS

              ( CUSTOMER
                JOURNEY )

        SALES                 MARKETING
```

→ R&D PROJECT TIMELINES ARE CRITICAL
   TO NEW PRODUCT SALES

→ A COMPANY NEEDS A PIPELINE OF LEADS

→ SALES TAKES TRAINING AND DISCIPLINE

→ CUSTOMER SUCCESS KEEPS THE CUSTOMER
   HAPPY, RENEWING, AND GROWING

GOOD FORECASTING TAKES A TEAM.

CHAPTER 4

The Foundation

Less than two days until the one-on-one meeting Sal asked for, Danny wondered what those ideas were that Sal mentioned.

Danny knew for sure he had better tighten up the R&D expenses and new product timeline. But what else?

They hadn't spent much time on any of the expenses in the meeting. *That's probably what Sal wants to talk about*, he thought. The expenses were just carried over from the previous year. Danny knew it couldn't be that easy. At least from what his uncle Bill was saying.

Sis hadn't had any new sales come in since Danny had been back home helping out. Cash was getting even tighter for Red Line.

Going through all the expenses, it became more obvious than ever to Danny that salaries were what was killing Red Line. *Wow, they pay out a lot every month*, he thought. Especially to the software engineers. "Well, Uncle Bill did say it can get expensive real fast," Danny said out loud, still trying to figure out what Sal wanted to talk about.

Then, he thought he had it. He and Sis needed to cut down the R&D team to reduce engineering expenses. That was it! That was the idea Sal was going to give him Thursday. Danny thought Sal must have seen that too and just didn't want to say anything in the meeting.

Danny knew this was a big decision and it was going to be tough. Some of the team had been with Red Line for years. Come Thursday, he'd be ready. Danny aimed to have a new, reduced staffing plan ready to show Sal. Danny's confidence was high.

When Thursday came, Danny was prepared. With the new staffing plan in hand, he walked into Sal's office confidently to tell him the news.

They sat down at the small conference table in Sal's office this time, instead of at his desk. No boardroom today. Danny assumed Sal wanted him to feel more comfortable than when he was across the desk from him. The table felt more casual. Like they were just having a conversation.

Sal started. "I'll be up front with you, Danny—you are showing a great deal of passion for learning the financial and business acumen you will need to get the company on track. But after seeing your forecast last week, there is still quite a bit of work to do. We didn't push on this too hard in the meeting last Thursday, but Phillip and I are concerned about your expenses in the forecast."

Perfect, Danny thought. *Sal thinks we need to cut. Here it comes.* He just knew Sal was going to be impressed.

"What we see, Danny, is that your expenses are too low. Especially around R&D. Properly forecasting the staffing needs for a small technology business like yours can make the difference between growing and imploding.

"We see a lot of small tech companies fail when cash gets tight because they forget how important it is to keep investing in innovation. The first place they try to cut is the engineering team. When the company is built on technology and new product development cycles, that is rarely the best answer.

"Phillip also took a quick look at some of the competition. Technology is changing fast. Advancement in AI has started to cut into the laser market, Danny. Would you consider doubling your R&D spend to get your new products out faster?"

Seriously? Danny thought. How could he have misread things that bad?

Danny was just starting to figure this all out. "I thought for sure you would want to cut expenses to get more cash flowing so we could cover the costs our family accountant pointed out."

"Yes, I can tell you are disappointed. This is where finances start to get a little confusing. Cutting costs can make you look more profitable at first, but it does not necessarily lead to more cash in the long term. That's what concerns your accountant the most, right? Cash flow?"

"Yeah, but..." Danny bit his tongue.

"Looks like you have an org chart in front of you, Danny."

"Uh, yes, that's right." This was getting uncomfortable for Danny. He knew he had made a major mistake.

"Great, let's take a quick look. How many of these people have you met with in building your forecast?"

Danny was feeling very concerned now. *Oh no. This is not going well.* He and Sis did the whole thing themselves.

"I know a few—most, actually—from past company events when I was younger. None were too involved in the forecast, though. Why do you ask?"

"Well, staffing decisions are serious stuff, and the expenses of a technology company like yours are primarily people related.

"Something I didn't mention before about *By the Numbers* is another requirement to be in the book. That is, to understand it's the people that make the numbers work. They are the foundation on which we build the business.

"For instance, you met with Julie over at Winning Robotics?"

"Yeah, she was amazing, Sal. She helped us a ton, especially with understanding the format for a good forecast."

"Glad to hear that worked out. Julie is amazing. She also hires amazing people. She knows people are the key to beating the numbers in the forecast.

"Danny, what would you think about revising your staffing plan? The staffing plan is directly tied to your sales and marketing budget as well as the R&D team we were just talking about.

"I want to encourage you to think about growth. Not just growth in the revenues, but growth in the technology too. And remember, people are the foundation to every business," Sal said.

"Yeah, that sounds like a good next step," Danny said, finally starting to feel a bit more comfortable.

"One more thing. I asked Phillip to give you a hand on your forecast."

"Sure, that would be great!" Danny said. "He seems really smart."

"Yes, he is. He has a lot of experience with technology companies. Phillip oversees all our technology investments. He can be a big help in getting your forecast right. Check in with him on your way out today. He is expecting you to stop by."

After the meeting with Sal wrapped up, Danny and Phillip met briefly in the lobby at Sal's office. Phillip told Danny he had set up a shared drive for where they could go back and forth with files and questions. Phillip also said he put several forecast templates in the shared drive to help get them started. Danny accurately sensed Phillip truly wanted to help Red Line be successful.

On the way back to the office, Danny gave Sis a call.

"Glad you called, Danny. I think we need to talk. The cuts you are suggesting to the R&D team are making me uncomfortable. We need to grow, not shrink, and we need those engineers to help us stay innovative."

Danny was confused. *What?* Did Sis know Sal was saying the exact same thing?

"Another thing, Danny, you need to get out there and talk to some customers! They need us so they can be more efficient in their operations. I know what we do is not glamorous, or cool like one of your dorm room startup buddies, but people need public transportation to get to work or school. For some people, this is the only way they can get their groceries. What we do is important!"

"Hang on, Sis! Where is this coming from?" Danny asked, almost in shock.

"The other day when we met with Julie, she said something that resonated with me. She said we won't achieve our vision if we stay in the red forever. We do have a vision for the company, Danny, and it's an important one."

"Okay, okay. I'll be there in a little bit, and we can sit down and go over it all together." That's when it dawned on Danny that Sis had been holding the business together while he was off at college. She had been the one in the trenches. Not him. The pressure had been building up.

THE FORECAST

***

When Danny walked into the conference room back at the Red Line office, he wasn't expecting what came next. Sis had gathered a few of the Red Line employees and was going through a spreadsheet she had projected up on an old big-screen TV.

He thought at first it was an intervention or something, but they hardly stopped talking to even acknowledge Danny had walked in.

Tommy T. was there. Michelle, the head of tech support. Mike was there too. He was one of the other sales reps who worked with Sis. Mike was always busy working on the website or posting on social media.

Danny sat down at the table, but before he could even figure out exactly what they were all looking at on the big screen, Sis asked, "Everybody know their action steps?" They all nodded and got up. It was like they had their marching orders and went out on their mission.

As Danny watched them walk out, he realized just how tight things were. He could see the stress in Sis's eyes. She wanted something better. And so did he.

"Sis, what was that all about?"

"We need to get better information about what is going on, so I enlisted some of the team to help. Tommy T. has some great ideas on how to streamline production and get more units out quicker, plus he has been experimenting with AI, which might actually allow us to increase price or sell some add-on functionality."

"Yeah, but what about the cost? Isn't that going to—"

Sis cut Danny off before he could finish. "Tommy T. has got this, Danny. He's sharp! And we can't afford to have him working on tech support issues. He needs to be driving our R&D."

Danny wondered if she had been talking to Uncle Bill. He said almost the exact same thing the other day.

"Michelle showed me some stats on all our tech support issues. It looks like one issue was causing us most of our support calls and even lost us a couple customers last year. She and Tommy T. are going to get right on it. Maybe we can get those customers back on board.

"The reason Mike was here is something I wanted to talk to you about. Mike has a real knack for generating new leads for us. Everything flows down from the leads and how they convert. Mike understands that better than anyone, and we need more leads. We can't afford to spend much on marketing, but writing high-quality articles, videos, and other content can make a difference for us."

"Yeah, but—"

She cut Danny off even quicker this time. "Danny, I would like to move Mike into a full-time marketing role, with the primary expectation that he will double the number of leads we get each month. He is excited about it. No more 'yeah, buts,' Danny. We are doing this!"

Danny heard her. Sis knew Mike well. They had worked together for years. If she thought he could handle marketing, then that was okay. Danny was feeling a bit humbled realizing he was still an outsider, new to the game, and not good at it yet either.

Sis was right. Uncle Bill said they needed input from a lot of people. Sal had also asked Danny who was involved. Danny assumed that must have been Sal's way of politely nudging him.

Before Danny could argue with Sis and say something he knew he would regret, a text came in. It was Ted. Perfect timing!

CHAPTER 5

The Milestones

The next day, the parking lot was full at Johnny's Diner. This was the place where Ted loved to meet up for coffee. He said it reminded him of when he and Danny were kids and would ride their bikes there for a burger and fries on Saturdays. It was one of those places that you couldn't believe was still in business at all but was always packed, usually with older couples who were retired and pinching pennies. Or maybe they just liked the homemade pies. Definitely a greasy spoon. Red vinyl seat covers and chrome on everything in an attempt to replicate a fifties diner. There were even posters of James Dean and Marilyn Monroe on one of the walls.

Danny almost always beat Ted there when they met. Ted was typically late and in a hurry like his company would immediately collapse if he didn't make every decision.

Today was no different. Ted was at least fifteen minutes late. Danny used to text him or wonder if maybe he messed up the time. Now Danny just waited, knowing Ted would eventually show up. The waitress glanced at Danny from across the diner like he'd been stood up on a blind date.

He tried to refocus on Red Line and what Sis was doing and how he could best help. She was clearly taking the reins of the day-to-day operations. For some reason, Danny had thought maybe that was going to be his role once they sorted out the cash problem, and she could go back to selling. She seemed so natural at the operations, though. The way she was working with the team back at the office was like the leaders you read about in those airport business books. Danny didn't know where he would end up. He was starting to be okay with that.

Just then Ted plopped down across the table. "Whew, sorry I'm late, Danny. Busy day! Back-to-back calls all morning."

"Good to see you, man! It's been a couple weeks—what have you been up to?" Danny said, trying to get the ball rolling so Ted could start his routine complaining session.

"What have I been up to? What have you been up to, Danny? Did you ever connect with Mr. Larobie? How did it go? Did he give you any money? I want to hear all about it."

A lot had happened in the last couple of weeks since Ted had given Danny Sal's number. Danny thought for a minute about where to start. He remembered *not* to say anything about the book. Sal said not to talk to Ted about it yet.

"Ted, I've got to tell you, it has been awesome meeting Sal—uh, I mean, Mr. Larobie. He has helped us so much. We are starting to get a forecast built out, and Sis—man, Sis is taking on the business. She is pulling the team together and—"

Ted cut him off. "Whoa, just a minute there, Danny boy. Let's back it up a bit. What do you mean Sis is taking on the business? That's your gig, Danny. Don't let her cut in on you like that!"

Danny knew Ted was trying to help and defend him. It did feel a little insulting to Sis, though, and Danny knew she was doing great. Danny had accepted she was certainly better with the team than he was. She knew the business. He could tell from watching the team respond to her in the conference room yesterday.

Ted had a way of rubbing people the wrong way. Including Danny. Ted was aggressive and made quick decisions. They weren't always the right decisions either.

"She's not cutting in on me, Ted. We are doing this together."

"So, what's your role in all this, Danny? Are you going to be her right-hand man?"

"Not sure that matters right now, Ted. It doesn't bother me not knowing either." Those words just came out of Danny's mouth without him even thinking. It did kind of bother him not knowing where he would end up.

"What? You're not concerned? Business is cutthroat, Danny. I know you're new to all this. Let me tell you how it works. If you don't take control now, you won't get another chance."

Danny wasn't going to let Ted get to him this time. "Ted, so how's that approach working for you? Last time we talked, you were nearly begging for a handout from Sal."

"Calling him Sal now? Guess your meeting went better than mine did. How much money did you get out of him?"

"It's not like that, Ted. We are a long way from determining if an investment would be the best move. Right now, we are meeting with tech leaders in the area and building out a forecast. It is helping me to learn not just our business, but business in general."

"Larobie tried to get me to go down that path once last year. Said he would have his driver help me build a forecast. Wanted me to go to some workshop or something. That pushed me over the top. I can run my own business, thank you. These money guys are all alike. They think they can tell you what to do. Once they get a piece of you, they won't let go."

"Ted, those 'money guys' are helping us. Not with money, but with connections, real advice, learning from their mistakes. Come on! Learn to open up."

"No way! Don't be a sucker, Danny. They will take advantage of you, and so will your sister. Wake up! The business world is a cruel place."

Danny and Ted had had fights before, but Danny couldn't remember any of them being this personal. "This isn't going anywhere, Ted. Let's change the subject. How is your business doing?"

"Not good. That's what I'm telling you, Danny. Life's tough. If it wasn't for me holding it all together, the business would fall apart completely."

"What is it this time, Ted? Did you lose a big customer or something?"

"Worse. My sales manager, Jessie, quit today. She said she thought there were other companies that needed her help growing more than I did. There could have been more growth here too, if she had sold more."

"Are you sure that's what it was?" Danny fully suspected she left because she couldn't stand Ted's leadership style anymore. "It probably wasn't just money, was it?"

"She talked about training, discipline, and process, but she has said that stuff before. She knows we can't do any of that stuff if we don't grow. She needed to sell more."

"So, what are you going to do now?"

"Guess I'll go back to selling. I seem to be the only one who wants to see this company win."

Danny thought he should say something encouraging. "Ted, you've got this. You've had to pull it together before; you can do it again. Keep your chin up."

"Look at you, Danny. Being all professional and adult. Thanks . . . I guess."

They went on to talk about football for a while before doing the bro hug shoulder bump handshake they had been doing since junior high. "Text me if you need anything, Ted," Danny said, thinking it might be more positive next time.

"Yeah, sure," Ted replied like he didn't think there was anything Danny could help with. It didn't bother Danny at all, though, because he knew Ted didn't think anyone could help him.

THE MILESTONES

***

When Danny got back to Red Line, he went straight into Sis's office. It was ridiculously well organized. Her desk was covered, but everything had its place. Phone, tablet, and laptop all matching in her favorite pastel blue and laid out in perfect order on her desk. Danny had always called her a neat freak. Her ink pens all lined up in a row. Not a speck of dust anywhere.

"Sis, I want you to know how great a job you are doing with the team. You have stepped up, and it shows." Meeting with Ted made Danny realize she might need encouragement too.

"Thanks, Danny. That was nice. It takes a team, and we have a good one. How did your coffee go with Ted?"

"Terrible. His leadership style is just . . . off. Something doesn't click. He doesn't listen. He thinks he has all the answers. He loathes investors like Sal and Phillip."

"Well, that's his loss. Our whole outlook on business and growth has changed since we met those guys. But I'm swamped with helping Mike and Michelle right now. Do you think you can tighten up the marketing piece of the forecast so we can have something to plug Mike's new numbers into?"

Danny was surprised. *Wait a second*, he thought. *What just happened?* "Uh, sure! That would be great!" Maybe that was where he could best help. "Hey, Sis, I was thinking on the drive back over to the office. Maybe we should add some narrative to our forecast about all the things you are working on with the team. The forecast would make more sense then."

"That's a great idea. Did Sal or Uncle Bill suggest that to you?"

"Nope—well, sort of, but not really. It just seems like part of what it takes to make the forecast real. The team won't buy into it if it's not realistic. Right?"

"Exactly! I'll ask each of them to write up a paragraph or two explaining their plan and how it will impact the numbers," Sis said.

"Perfect. I'll go grab the book and see what I can learn about forecasting for marketing," Danny said as he left Sis's office.

He remembered from the call with Julie that he had better be ready with his questions.

He found the book in the conference room and looked up *marketing*. Now what was that line again? "I have a question about the numbers. Will you help me?"

Ping. Sent . . .

Nothing . . .

Danny went back into Sis's office to tell her it wasn't working. She told him to relax and give it some time and stop acting like Ted.

Danny decided to take the time to go talk to Tommy T. about timelines. It was the first time he had been in the R&D lab in years. It wasn't fancy like Uncle Bill's, or cool like Julie's. Only a couple of developers; there were a few empty desks from when the company was bigger. Sis had told Danny their dad held off as long as he could but finally had to make some cuts.

"Hey, Tommy T.! Uncle Bill asked how you were doing the other day," Danny said as he walked in.

"Nice. Haven't seen him in years. What's he up to?"

"He's working on some new AI image recognition project. You should give him a call. Sounded right up your alley."

"Will do. We can compare notes. We have one on the back burner right now too."

"Oh yeah? That's what I wanted to talk to you about. I'm working on this forecast with Sis, and Uncle Bill was telling me about how they do forecasting over at his office. He says they start with getting feedback and timelines from the R&D team."

"Of course, he would say that. They have the luxury of having tons of developers that can crank out new products whenever they want. He hated being in a small team like ours. He always said we would be behind the curve. I think that's why he left us," Tommy T. said.

"Yeah, I kinda got the same feeling. He really seems to like the big team they have over there."

"But that's okay, Danny. We don't have a big team like he has, but we do have a good development process. Let me show you what all we have going on."

They walked over to the big chart on the wall. The entire wall was covered with vinyl whiteboard wallpaper. It had to be at least fifteen feet wide. There were all kinds of diagrams and arrows pointing at boxes and circles.

"The part you're looking for is over here. It's a rolled-up project timeline of all our main projects. Sis knows all about them. That's what we were talking about in the conference room when you walked in on us the other day."

"Okay, so what does this mean for our forecast?"

THE MILESTONES

"You can see it is broken down by month from left to right, with each main project listed down the wall. For instance, we are working on a chat tool for Michelle to reduce customer tech support calls. If we can stay focused, it will be done in less than a month. That dot is a completed milestone."

"What's the arrow on the 'defect fixes' line?"

DEVELOPMENT TIMELINE

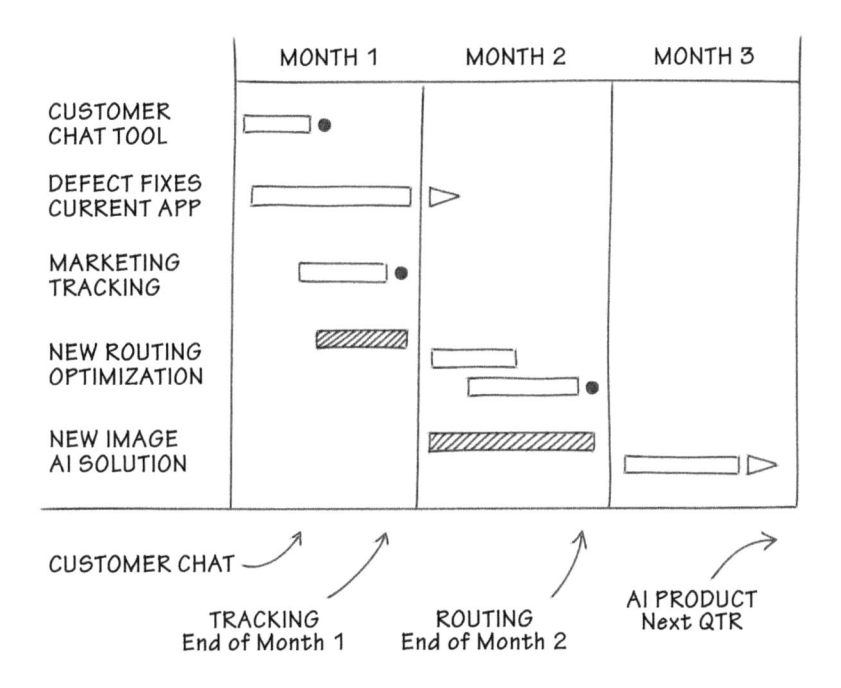

	MONTH 1	MONTH 2	MONTH 3
CUSTOMER CHAT TOOL	▭●		
DEFECT FIXES CURRENT APP	▭▭	▷	
MARKETING TRACKING	▭●		
NEW ROUTING OPTIMIZATION	▨	▭▭●	
NEW IMAGE AI SOLUTION		▨	▭▷

CUSTOMER CHAT ↗
TRACKING
End of Month 1

ROUTING
End of Month 2

AI PRODUCT
Next QTR

63

"That means the project will be ongoing. There are always going to be issues we need to address. We try to build that into every project."

"The other day, Sis and I presented our first pass of the forecast to Sal Larobie—"

"Really?" Tommy T. cut Danny off before he could finish. "We pitched him years ago. Your dad and I wanted to go it alone, but I think your uncle Bill wanted us to take his offer. Are you and Sis looking for money?"

"No, but we know we need help. And he is connecting us around town. We just want to get this forecast right so we can get his feedback on it."

"Okay, I got it. Then the projects you are looking for are down at the bottom. The new routing optimization might give Sis a way to increase prices. The very bottom one, the AI image recognition, is the big one. It won't be done this quarter, but it could increase who all we can sell to and save some of the customers we lose."

"Yeah, that's what I'm looking for. The routing has a dot, a milestone, at the end of month two. That's when we can start selling it, right?"

"Yes, that's right. It's okay for marketing to get started with their prep work sooner. But your dad would never sell anything until it was completely tested and ready to go to the customers. He was very serious about that. I hope that doesn't change either."

"I understand. Will you look over the next version of the forecast before it's finished to make sure the dates match what you want?"

"Absolutely, Danny! I would love to work with you and Sis on that."

About that time, Danny's phone went off with a call coming in.

"Hi, this is Eric with Stream View Digital. Just saw your text from earlier this afternoon. How can I help you today?"

TOMMY T.
RED LINE R&D LAB

MEETING NOTES

HAVING A <u>GOOD DEVELOPMENT PROCESS</u> CAN MAKE UP FOR NOT HAVING A BIG TEAM.

MILESTONES = BIG TIMELINE EVENTS

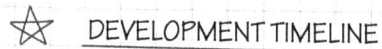

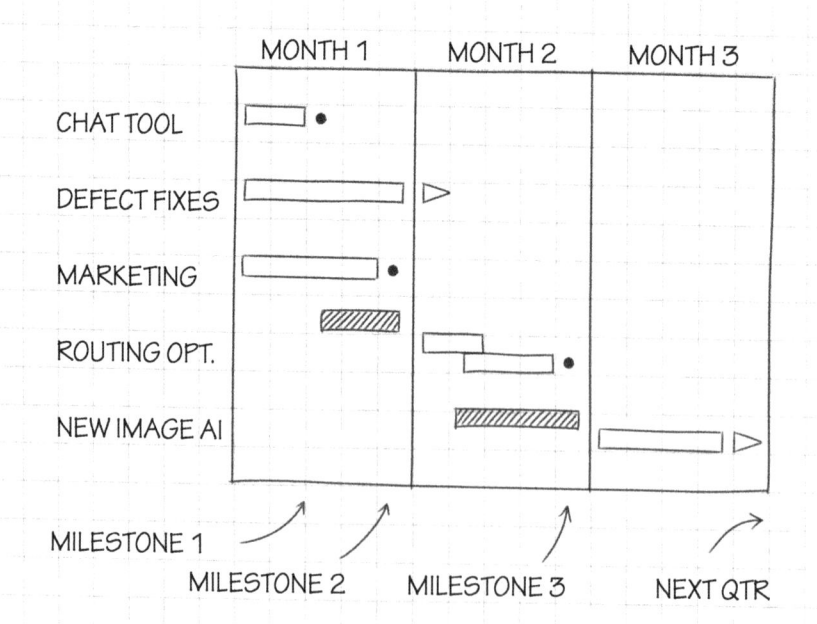

☆ <u>DEVELOPMENT TIMELINE</u>

	MONTH 1	MONTH 2	MONTH 3
CHAT TOOL	▭ •		
DEFECT FIXES	▭▭▭	▷	
MARKETING	▭▭ •		
ROUTING OPT.	▨▨	▭▭ •	
NEW IMAGE AI		▨▨▨	▭ ▷

↑ MILESTONE 1
    ↑ MILESTONE 2
        ↑ MILESTONE 3
            ↗ NEXT QTR

→ EARLY PREP WORK WITH MARKETING

→ SALES HAS TO WAIT UNTIL EVERYTHING IS TESTED, ACCEPTED, AND READY TO GO.

DON'T SELL UNTIL THE PRODUCT IS READY!!!

CHAPTER 6

The Funnel

"Hi, Eric, this is Danny over at Red Line Technologies. I need to learn about how to forecast our marketing leads for new sales."

"Hi, Danny, good to meet you. Great topic! Happy to help. Want to get together to talk about it?"

"Yeah, if you don't mind. That would be great!"

"I'm driving in from the airport, and traffic is a little backed up, so it will take a while to get to the office. How about six this evening? Is that too late? The office should be cleared out, and we can have some peace and quiet to talk."

"Perfect, Eric! Thanks for being so open to help. And on a Friday evening too!"

Eric met Danny at the door when he arrived at the Stream View office. It was an old brick military building that had been turned into upscale offices. Lots of glass windows, concrete floors, and metal beams everywhere. The ceilings were all exposed, so you could see the ductwork. Like a warehouse. Danny assumed it was done just for looks. Everything was obviously new. Only the brick shell was original. It was cool, Danny thought. Probably expensive. Danny found himself thinking about the cost of things much more often since being home.

The Stream View offices were on the second floor. As Danny and Eric walked up the open stairs from the main floor, Danny could see down into the lobby where they had just left. The bright blue and green modern chairs and stools were scattered around where people had used them throughout the day. For a second, Danny envisioned all the conversations happening in small groups, chatting, collaborating, pointing at screens on their laptops. He thought it was surely a busy place during the day.

As they went through the Stream View double glass doors into the reception area, Eric stopped. "Danny, I want you to know, earlier today, I was on a flight coming back from a quarterly business review with a customer. That's why it took me so long to get back to you."

"No worries, Eric. I'm just glad you were able to meet today."

"When someone uses Sal's *By the Numbers* book, I know it is serious and they need help. So, I always try to get back to them ASAP. It is a powerful system to be able to learn and help others. I use the book too. Probably more than I would like to admit. It comes in handy."

"Really? I just assumed everyone in the book already had it all figured out."

"No way! There is so much to it and the topics are so intermingled, we will never have it all figured out. We just take one learning at a time and see how it fits together. Anyway, I'll give you a tour in a little bit, but first tell me about Red Line Technologies. Then we can get to your marketing questions."

Danny could tell Eric was polished and had spent a lot of time in front of audiences. Lots of eye contact and an inviting smile.

"Sure, Eric. I love talking about Red Line!"

That was the first time Danny could remember saying that. Maybe it was the first time he had felt that way. Talking with Tommy T. got him excited about the new tech on the horizon.

Danny had heard his dad and Sis do this pitch enough over the years. He thought he should be able to pull it off. *Here goes nothing*, he said to himself as he dove in.

"It's simple. People drive our business, and we make technology to help drive them. People need to get from place to place every day. It could be school, work, or even shopping for groceries. We work to make local public transportation more effective and efficient. We do that by using data analytics and deep learning from our laser technology people counters on public buses and trains, all over the US."

Ugh. That needs some work, Danny thought. He liked the people angle. And data analytics. The laser thing sounded old and out of touch. Oh well, he thought. He got it out there. His first elevator pitch. Yay!

"Danny, that is great that you have such a clear vision and passion for helping people get from point A to point B faster and cheaper. I remember reading something about your company years ago in the local tech journal."

"Could be. My dad used to hype it up a bit around here. He always said it's good to get press, even if it's local."

"Does your dad still run the business?"

"No, he passed recently and Sis, uh, my sister, Cecily, and I are partnering to get things back on track. She is doing great. She knows the business, the team, our customers. It is enjoyable to watch her work."

"Sorry to hear that about your dad. I'll bet you and Cecily have it under control, though."

"Not exactly. That's why I texted you. We hit a wall on growth, and we need to better understand how to forecast our leads from all our marketing work."

"Okay, tell me a little about what you have so far for forecasting marketing."

"Well, we presented our first draft to Sal and Phillip the other day, but we focused mainly on new sales, and not so much on marketing. Now we realize that was a gap."

"Phillip! Great guy! He is a good man to have in your corner. We went through an acquisition last year where we bought a smaller agency, with Sal's help, of course, and Phillip coached us through the entire process, beginning to end. He's a pro! And you're right, Phillip will definitely catch any gaps in your forecast. So . . . how do you get your new customer leads today? Do you run a product-led model or a sales-led model?"

"I'm not sure what you mean. We have salespeople, if that's what you're asking."

"Sort of, but not exactly. A product-led model, or some people say product-led motion, is where the product does not require much human intervention to land a new customer. For instance, a prospect might go to your website and sign up for a version of your software. Then, if they like it, buy it, upgrade, and so forth."

"Oh, that makes sense—but, no, that is not us. I mean, we have a website, and Sis—uh, Cecily—says we have good SEO, but the product does not sell itself. We have to do demos, and sometimes pilots where the customer gets to actually see it in action. We also have a hardware piece."

"Oh, that's right. The lasers."

"Yeah, but we are building some new AI tech that might eliminate all the lasers."

"Okay, I get it—then you have a sales-led model. No problem. That is common for enterprise-level software sales. Do you know where most of the leads come from today? You mentioned SEO."

"Yeah, that and trade shows. We go to several trade shows every year. We also get referrals. Sometimes when a user leaves one of our customers, like a city transit authority, and moves to a new town, they might recommend us. That happens."

"All right—so far, so good. Do you have a marketing team at Red Line?"

"No, we only have a sales team. But one of our sales guys has been handling the website and social media and is now going to move full time into marketing."

"Good to hear. Marketing and sales truly are two different areas, each with their own business metrics and specific skill sets. A lot of companies lump them together. We did too, for a long time until we were large enough to staff both properly."

"Yeah, that's where we are right now," Danny added, like he knew what he was talking about.

Eric went on, "Marketing is all about positioning the company in the market so prospective customers understand how you can help them. Marketing comes down to generating high-quality leads. It's not just the number of prospects but making sure they are the right prospects too."

"So, what's a quality lead?"

"It's a potential customer that aligns well with the products you sell and at the price you sell them. Sometimes you might get leads through your website that are all wrong. They are looking for something completely different than what you offer. Those are low-quality leads because your products don't align well with their needs. It's not that they are low-quality companies. Just that their conversion rates would be low when moving them through this marketing funnel. Compared to high-quality leads, which would have a high conversion rate."

"Hang on, what do you mean by marketing funnel?" Danny asked.

"Excellent question. I didn't know if that would be new to you or not. The marketing funnel is something I love to talk about. It is even something I've taught in one of Sal and Phillip's workshops. Here, let me draw a picture of how it works."

By then, Danny and Eric were down the hall quite a bit from the front lobby. There were lots of little breakout areas with rolling whiteboards. Eric grabbed one and rolled it over to where they were standing.

He continued, "Let's start by listing all the ways a prospect might hear about your product."

"Okay, uh, the trade shows I mentioned. And the search engines for sure. Oh, and we run ads in a couple city-focused newsletters too."

Eric was writing as fast as Danny was talking. Then he asked, "What about social media? You mentioned one of your salespeople is handling that. Does it generate any leads for you?"

"No clue. I mean, I wish I knew, but we just haven't focused much on tracking this stuff. At least that I know of."

"Well, you might be surprised. Sometimes there is already tracking in place you can tap into. If not, it is not too hard to have tracking set up for all this. I'll go ahead and add a couple others. Let's add paid search for the search engines, and review sites where people can see your solution ranked against the competition. Those are always good for enterprise software."

"Now I'm getting it. All these different lead sources are what go into the top of the funnel. Right?"

"Yes, that's it exactly. Now, as the prospect goes through the funnel, they also go through phases. At first, they are just becoming aware of your product or company. They might be searching for a solution or doing a little research. Then, once they find you, they might show interest and want to learn more. They will start evaluating your solution, maybe compare it to others, and maybe consider alternatives."

"So, we need to get more into the top of the funnel?" Danny asked.

MARKETING FUNNEL

```
         PAID SEARCH
SOCIAL MEDIA    TRADESHOWS
BLOG SEO              REVIEW SITES
REFERRALS             NEWSLETTER
```

- Becoming Aware, Searching, Researching
- Showing Interest, Learning, Considering, Evaluating
- Converting, Qualifying

Track #, %, Time, Source

"Yes, but not necessarily. It is what comes out of the bottom of the funnel that matters. For each phase the prospect goes through while in the funnel, there are things you can do to help them in their journey."

"The customer journey?"

"That's exactly right. The customer, or prospect in this case, is on a journey to find a solution to their problem, and you can help them by providing the right information at the right time. It might be a PDF of a whitepaper or case study about how your technology works. Or it might be a short video. There are lots of techniques to help the prospect with their decision-making. It is critical to the success of a marketing program to be able to track their journey. To know where they came from, what they downloaded. Did they open an email from your team? Did they read it? All that information will help you forecast your number of leads."

"Eric, this is awesome. We gotta make sure we have tracking in place!"

"That's right. Another day we can talk about how to get more leads. That's what we do here at Stream View for our customers. But for today, and for your forecast, just try to create a set of assumptions for each source that shows the numbers coming in, the percentage making it through the funnel, and how long it takes for them to get through. That will bring life to your forecast. Now, remember how I mentioned I had just come back from a customer quarterly business review?"

"Yeah, I was going to ask you about that. Before we dive into that, give me a minute to catch up on my notes."

ERIC, CEO AT STREAM VIEW
STREAM VIEW DIGITAL OFFICE

MEETING NOTES

PRODUCT-LED VS. SALES-LED MODELS
- PRODUCT-LED IS SELF-SERVE
- SALES-LED REQUIRES PEOPLE

→ MARKETING + SALES ARE DIFFERENT
    MARKETING — POSITIONING + GETTING LEADS
    SALES — CLOSING DEALS

MARKETING FUNNEL PHASES

1. AWARENESS, SEARCHING
2. CONSIDERING, EVALUATING
3. CONVERTING, QUALIFYING

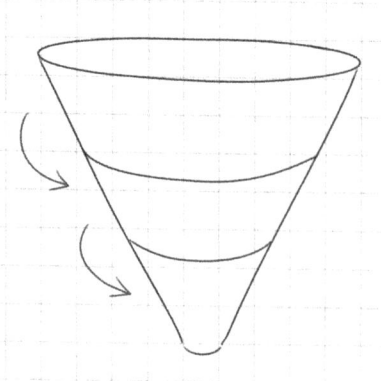

☆ TRACK EVERYTHING!
    NUMBERS, PERCENTAGES, TIME, SOURCE

NOT JUST MORE LEADS — THE (RIGHT) LEADS.

CHAPTER 7

The Customer

"Thanks. Okay, I've gotta keep my notes up to date. There is a lot to learn. Where were we?"

"No problem. I'm glad to see you taking it all in. We call it a quarterly business review, or QBR. It is where once a quarter we meet with our customers to talk about their business, and how we can better help them solve problems. Sometimes it is face-to-face like the one this week, and sometimes it's a video call. But it must happen every quarter, no exceptions. I personally try to meet with as many customers as possible at least once per year. Our team handles most of the meetings, though."

"That sounds like a lot of work and a lot of travel, isn't it?"

"Yes, it can be intensive. It is one of the ways we stay up to date and make sure we add value. The QBR makes sure we are delivering on what we promised. We focus on the customer's team and how we work together. What can we do better? We ask for feedback on our people and our services. Here, this is what I wanted to show you. See down this wall, all these two-foot-by-three-foot posters?" Eric pointed.

"Yeah, that's neat to have all those customer logos up there."

"These are what we take to our QBRs to show our customers. The charts on these posters are the improvements in marketing efforts we have made for the customer during the last quarter. It's like an enlarged version of a marketing dashboard."

"What a view of all your marketing in one place! That way they know you're good at what you do."

"Sort of, yes. Sometimes there are some celebrations and high-fives when working with our customers. But sometimes the charts don't look that great, to be honest. That's when we know we must buckle down and work together with the customer to fix things."

"Ever have a customer fire you when things don't look good?"

"Not fire us exactly, but yes, we are not always the best fit for everyone. And occasionally we part ways with a customer. However, these QBRs help everyone stay on the same page so there are never any surprises. We always know where we stand. The main thing to remember today, Danny, is to always be listening to the prospects and the customers. If you want to get a real view of your business, ask your customers. That's how you can innovate."

By then Danny and Eric had made it down the hallway to the snack bar. It was a contemporary layout with artsy tables and chairs. What caught Danny's eye, though, was the amazing view. "Wow. Now this is the real view! This is inspiring."

"You can see the mountain summit from here, and all the streams and rivers coming down. That's why we named the company Stream View. It's good to have inspiration to keep your vision focused.

"Danny, I would like to help you with connecting the feedback from your customers to your marketing strategies. We find our QBR process is the best way to keep our marketing aligned. Would you be open to me sitting in with you and your team?"

"Absolutely! That would be great. Eric, you have been a huge help today. Thanks for taking so much time!"

On the way out, while Danny and Eric were exchanging small talk about the office and how Eric grew his company, Danny realized the people helping him and Sis were not asking for anything in return. They all went out of their way to help.

It was almost eight and pitch black outside. Danny had overstayed his welcome. Eric never said a word about it, though.

As Danny got to his car, he started talking to himself. "That's who I want to be. Someday, I hope to be able to help somebody else learn this stuff too, like everyone is helping Sis and me."

Before he pulled out of the parking lot at Eric's, he texted Sis. *Hey, Sis, had a great meeting with Eric, the founder at Stream View Digital. He had lots of marketing ideas for us. He suggested we do some customer QBRs. He also offered to meet with us to go over his process. Let's talk about it first thing Monday.*

***

Monday morning when Danny got to the office, Sis and Michelle had the existing customer list up on the old TV in the conference room. Sis always beat Danny to the office. She beat everyone to the office.

Danny told them all about the meeting with Eric and how he had offered to come over and meet with Red Line. Danny thought they would be excited. But Sis and Michelle looked worried.

One of the customers on the list was a problem. Most of the others were either in good shape or fell into the group Michelle talked about the other day when she said there were some defects needing to be addressed. That was not the problem today, though.

Michelle went on to explain, "Like I was tellin' Sis before you came in, Danny, your dad and I had been workin' with this customer last year. They made a large up-front payment but then claimed we misled them about how our system worked. They thought it already had all the AI tech that Tommy T. is still buildin'. We think they got us confused with a competitor who has AI already. They want their up-front payment back. They even threatened legal action. I thought it was handled. Guess it's not. This is bad news, Danny, really bad."

Sis jumped in. "Let's not get ahead of ourselves here, Michelle. It may not be that bad."

Danny could see the reassurance Sis was looking for, so he backed her up. "Yeah, Sis, that's right. Maybe this is an opportunity to meet with them and hear what they have to say. Kind of a learning experience."

"You don't want that meetin', Danny. I've met with them before. I think they liked me, but they weren't happy. I thought this was settled already. I don't wanna drag it all back up. They made your dad mad. He thought they were callin' him a liar, and then he walked out of the meeting."

Danny's head was spinning at this point. Not only were there financial problems, but there was also a legal problem. Customers were mad. Technology was old. Bugs needed to be fixed. But Danny knew his dad was not a liar.

Sis jumped in. "We must meet with them, Michelle. It won't be easy. I will call them and set something up. Can you get me caught up on the history and find their contact information? We will show them how much we want to solve the problem."

Once Michelle left to go dig up the history, Sis slumped down in the chair next to Danny's. "I'm getting tired of all this. Up and down, day after day. As soon as we make progress, something else pops up."

"I know. Friday night I was all excited again after meeting with Eric. Every time I meet with someone outside of Red Line, I get excited. Then I come back down to earth and realize what we are up against."

"It's overwhelming," Sis added.

Danny and Sis sat together in near silence for several minutes.

Then Danny walked over to the whiteboard. "Sis, we can do this. We already have a plan. Let's go through it together."

"What do you mean, Danny? We don't have a plan."

"Sure, we do. Remember those paragraphs you wrote up last week for the financial forecast?"

"Uh, those aren't done yet, Danny. Sorry, it's only been a few days."

"That's okay. Those paragraphs will fit perfectly with what Uncle Bill was telling me the other day about their forecasting process. They call it a customer journey. It's a circle that goes on forever where they get input from different groups in the company. Like Tommy T. for the project timelines, and Mike in marketing. Each group owns their part of the forecast."

"But Danny, we don't have a big team like Uncle Bill."

"We're closer than you think." Danny could hear Friday night's conversation with Eric coming back in his head about the QBRs and what customers thought about products and services.

"Next, we need to know how our customers value us. I've got an idea on that one. Friday night, Eric was talking about QBRs. You know, Michelle always seems very reactive. Not proactive," Danny said.

"Now wait a minute, Danny. Michelle wants to be proactive. She is just swamped with tech support calls," Sis said.

"Hang on. That's where I was going. I think she could head up a new initiative to start doing QBRs, like Eric. This could be a proactive process rather than reactive. It would let her get ahead of the issues before they become unmanageable like the one you two were talking about a while ago."

"Okay, I'm listening," Sis said.

"One of the things I've noticed more than anything else in meeting with people in Sal's book is how open they are. They seem to want to hear what we have to say. They listen. Ask questions. That's who I want to be too. And that's the kind of company I want us to be."

"Me too, Danny. But how do we get there from here? Right now, we've got real problems."

"Well, you showed me how to do it the other day—when you pulled Tommy T., Mike, and Michelle into this very room. You showed me that being open and receptive is what it takes. You almost yelled at me about it. Said I needed to get out there and meet some customers. That's exactly what Eric was saying too. Let's go get them to open up and talk to us about our people, how they are doing, our product, and what needs to change for it to be the best solution to their problem."

"The team is a little nervous about 'poking the bear,' Danny. They've already told me not to call on certain customers in the past when I was doing the sales."

"I don't think that's an option anymore, Sis. We must poke the bear to find this stuff out. We just have to. Eric says hearing the customer feedback is also what drives their marketing strategies. He's a marketing guy. He should know, right?"

"Okay, okay, I get it. You're right. How about if we put Michelle over all our existing customer relationships, with a goal of meeting with them each quarter and finding out what they truly think of us. Just like you said Eric does," Sis suggested.

"Okay, let's tell her right now. Before we change our minds. Then we get back to the legal issue we were dealing with. I'll go get her," Danny said as he headed out the door.

***

When Danny got to Michelle's desk, she looked both mad and sad at the same time.

Michelle was the real deal. She was sincere and loved the Red Line customers. Down-to-earth. Had a little bit of a southern accent that worked well with customers. Michelle had worked her way up through tech support in the company right out of high school.

"Danny, I'm sorry about this whole legal thingie. I should have told you and Sis about it sooner."

"Come on back to the conference room. Sis and I want to talk to you," Danny said, coming off a little too harsh.

Danny noticed Michelle looking over all his chicken scratches of Uncle Bill's customer journey circle on the whiteboard as he and Michelle walked back into the conference room. She looked worried as she sat down across the table.

"Guys, come on. We can work through this, right?" Michelle said.

"What do you mean? Of course we can. That's what we want to talk to you about," Sis said.

"Let's get this over with."

Sis jumped in. "Oh! No, you're not in trouble! We need you now more than ever."

"I'm not?"

Sis then added, "No, of course not. Actually, we want you to head up a new initiative to focus one hundred percent of your time on proactive customer outreach, instead of the reactive tech support you've been doing. Kind of a customer success initiative. Would you be up for that?"

"Heck yeah! I thought you were gonna fire me or something for that legal problem. I'll do whatever you need me to. I love this place."

"Perfect. We thought you would want this. We need to do some shuffling to handle the tech support. Any ideas on that?" Sis asked.

"Well, we haven't talked about it yet, but yes, I have some ideas. We've been testin' some chat tools internally. The answers we get are nearly one hundred percent right, but the tone of voice is still off. We are gettin' close, though."

Danny noted, "Yeah, Tommy T. has it in his project timeline—you know, the big chart on his wall?"

"Yeah, that's it. We're almost ready with it. Should be done this month. It's in his hands. I can tackle the proactive customer stuff anytime," Michelle said.

"Perfect. I'll set something up with Eric to go over his QBR process," Danny said.

"Hey, hold your horses. I got my own process I've been workin' on. Here, let me show you. It's nothin' fancy, but it works." Michelle grabbed the markers out of Danny's hand and jumped up to the whiteboard and drew four long arrows from left to right.

"It starts when we first get a customer from sales. We want to get more info and get it sooner too, by the way. Hint, hint." Michelle smiled and glanced over at Sis. It was good to see her confidence coming back.

"Yeah, I know. Sorry about that," Sis said.

"Then, after we onboard 'em on the software, we work with them to get them using it. We call it activation. We figured out getting them onboarded is not the same as getting them active. If we don't get 'em active fast, we kinda lose them along the way."

"How do you know if they're active?" Danny asked.

"Well, it's kinda tough right now. Tommy T. gives us some login stats—that helps. What we really want, though, has more to do with how much of the data tools they're using. That's when we know we have 'em locked in, and they're hooked on it. Then they love us."

CUSTOMER SUCCESS PROCESS

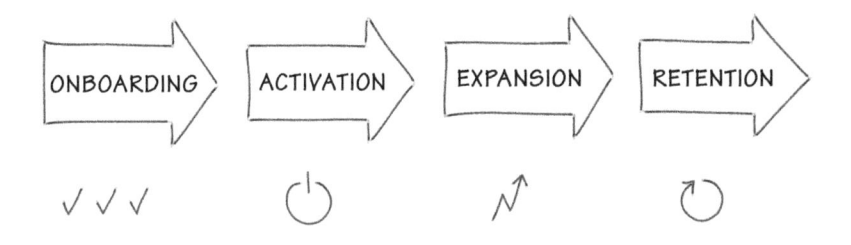

ONBOARDING → ACTIVATION → EXPANSION → RETENTION

✓ ✓ ✓   ⏻   ↗   ↻

"How do you know when they're hooked, Michelle?" Sis asked.

"Well, we hardly ever lose anybody once they start using the data. We only lose the ones that just take the laser reports without any analysis. For real. We can tell the difference just from the type of tech support questions they ask."

"You're serious?" Sis asked.

"You bet! I've been trackin' this for years. That's how we pick which bugs to work on. When Tommy T. is shorthanded, like he is now, we gotta pick and choose. So, we go with the bugs in the analytics stuff 'cause that's what our best customers use most."

"Sis, this is exactly the kind of customer intel we need. I'll bet this is what Eric was talking about," Danny said. "This is great, Michelle. What else?"

"This next step in the process is the area where I know our competitors try to grow their customers. We don't do much of that right now. What a customer pays when they start is pretty much what they pay us forever. I don't know how to get them to pay more. Then this last part is how we try to keep 'em. Retention, or something. It needs a better name. You know, keep 'em from canceling their contract. I hate it when somebody cancels. We got this spreadsheet with all the contract dates in it. If I just had more time . . ."

"Michelle, you've got this! We will make sure you have the time you need now. It is obvious you care about our customers. And it looks like you already have a good process. We can work together, and on your process too," Sis concluded.

"Awesome!" Michelle said as she handed Danny back the markers and smiled like she had just won a ribbon at the county fair.

Tommy T. walked in as Michelle strutted out.

MICHELLE + SIS
RED LINE CONFERENCE RM

MEETING NOTES

→ USE QUARTERLY BUSINESS REVIEWS
    QBR

→ DON'T BE AFRAID TO POKE THE BEAR

→ GET MORE INFO FROM SALES,
    AND <u>SOONER,</u> TOO

MICHELLE'S STEPS FOR CUSTOMER SUCCESS:

- ONBOARD
- ACTIVATE
- GROW
- RETAIN — <u>KEEP THEM HAPPY</u>

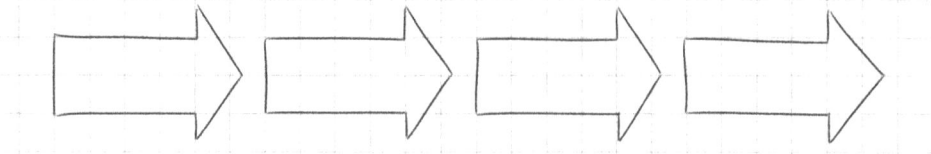

⇒ ⇒ ⇒ ⇒

☆ ONBOARDING IS NOT THE SAME AS ACTIVATION

GET THE CUSTOMER ACTIVE <u>FAST!!!</u>

CHAPTER 8

The Purpose

"Sorry to interrupt. I couldn't quickly find the electronic version of the inventory, but here is a hard copy from last year. Hope this helps," Tommy T. said.

"This is great! Thanks." Danny said as Tommy T. left to get back to work.

"What's that?" Sis asked.

"It's an inventory of the old lasers we don't use anymore. Tommy T. was telling me they are outdated, and we can't sell them. I was thinking maybe Julie over at Winning Robotics might be interested."

"Hey, that's a great idea," Sis said.

"Did you two ever connect about that leadership conference?"

"Not yet. I don't know. Never thought of myself as a leader."

"Sis, you should do it!" Danny said. "I'll go talk to Julie about the inventory, and I'll tell her you will call her about the conference."

"Uh, okay, I guess. It sounds kind of soft and fuzzy, though."

"Come on, you're great at this people stuff, Sis. I was impressed with how you brought the team together. They are listening to you. You know, all the people we've been meeting with—Julie, Uncle Bill, Eric, and especially Sal—they all say it is more than the numbers. It's the people that make the numbers happen. People are the foundation. You've got this, Sis!"

"Hang on a minute, Danny. There is something else," Sis responded. "It seems like we are still missing something. Dad had bigger ideas for the company. I mean, let's say we get all these numbers sorted out, and the team, and the customers, everything. Then what? Why are we doing this?"

"What do you mean, bigger ideas? Isn't what you said the other day about helping our customers be more efficient in their operations enough?"

"Not even close! Dad had big ideas. Maybe they got in his way, but at least he had a vision. A purpose for the business. He wanted to help communities become healthy.

He thought of public transportation as the veins and arteries of the community. He wanted our software to run everything. I remember him talking about how someday we would be able to leverage data to help reduce crime, to increase access to education, even eliminate homelessness!"

"Sis, you're getting carried away. Let's just try to run a profitable company and pay the bills. Okay?"

"No, Danny. We aren't leaving this room today until we add some of Dad's vision to the plan you have on the whiteboard!"

"You're serious?"

"Yep. No matter how profitable we can make the company, it doesn't matter if we don't have a purpose," she said.

"How about I go call Julie about the inventory and you can add the vision stuff to the whiteboard."

Sis was already at the board starting to write something about AI and optimizing bus routing.

He texted: "Hi, Julie, this is Danny over at Red Line. Have time for a call or a coffee this week?"

The phone rang. Julie said, "Hi, Danny. I was just thinking about you and Cecily. It's been a couple weeks. There is a demonstration session with the kids at a local elementary school today near your office. Want to join me?"

"Sure, that sounds like fun. What school and what time?"

"The presentation is at two this afternoon over at Mountainside School on—"

"Hey, that's where Sis and I went to school. Haven't been there in years. I would love to join you."

"Okay, can you be there at one thirty to help me set up?"

After Danny and Julie chatted for a few more minutes, Danny went down to see Tommy T. again about the inventory. "Hey, Tommy T., how ya doin'?" Danny said, carrying over a little of Michelle's country accent.

"What's up, Danny?" Tommy replied.

"I've been hanging out between Sis's office and the conference room now for nearly three weeks. I want to get out of her hair. Think I could use one of these empty desks out here with your tech team?"

"Sure, glad to have you."

"Perfect! Just need a place for my laptop! And maybe an extra monitor."

"You know, your dad would be proud to have you joining us, Danny. He always wanted you to be part of the company someday. Said you were a real strategic thinker," Tommy T. said.

"Really? I never got that impression from him. Strategic thinker, huh? That reminds me, ever think about getting out of the hardware business and focus on the AI and data analytics?"

"Are you kidding? I think about it all the time. The entire team hates the hardware piece. We built a prototype of an AI image recognition system that is producing far more accurate and faster results than our lasers. We can do a lot more with the data too. The world runs on data nowadays, Danny! The more we can get, the better. See, there you go, thinking strategically," Tommy T. said.

"Thanks. Oops, sorry to cut out, but I gotta go. I'm meeting Julie from Winning Robotics over at Mountainside School to sit in on a presentation she is doing for the kids there. Trying to get her to buy our old inventory too. Thanks for the list."

***

Danny beat Julie to the school and started to look around. There had been a lot of remodeling since he was there as a kid.

The hallways were a lot narrower than he remembered. Danny recalled how he and Ted used to get into all kinds of trouble there.

"Hi, Julie! Good to see you again," Danny said.

"You, too, Danny. Let's get rolling. These are always fun. We have the room they call the science lab. It is not very big, but it works for these presentations."

"I remember that room. It was always my favorite part of school."

"Yep, the kids love this stuff. You'll see. I love it too. Wish I could do these every day."

The kids were totally zoned in on everything Julie was showing them. How robots could walk and talk. Way more than Danny had ever seen before. The kids would type a word on the tablet, and the robot would say it back to them.

THE PURPOSE

It was the coolest thing, seeing Julie be so engaged in sharing her products, her vision, how she believed robots could help people who couldn't do things on their own anymore. She wasn't just teaching the kids; she was inspiring them. And inspiring Danny too.

After the class, while Danny and Julie were packing up and carrying things back out to her car, Danny mentioned how impressed he was with her vision.

"Danny, that's why I invited you to join me today. All this work you and Cecily are doing on your forecast has to roll up to something. Something greater. I wanted you to see that business is way more than just numbers. Sure, you must know the numbers, but we are all in business for a bigger purpose than making money. We make the money so we can accomplish something."

"So, what is it for you and Winning Robotics? Helping the kids? You were great in there today."

"We started out making robots for gaming competitions, then we added educational robots for computer science students," Julie said. "So, yeah, for the kids. That still wasn't our long-term vision, though. It was good for a while. We could tell we were doing something good for our community. What we always wanted to tackle was making the robots to help people with special needs."

"I thought you were mainly focused on the industrial robots like those agricultural crawlers and drones and stuff?"

"Yes, we do that too, and it is a very strong part of who we are. The reason we do it, though, is to keep growing our revenue streams and profitability so we can innovate new and better ways to build our human-assistance business for the people who need us most. All our industrial clients know our vision too. That is part of why they work with us. They want to achieve that vision with us. If they didn't, we probably wouldn't be a good fit for them."

"This all sounds a little altruistic. I thought businesses were built to make money?"

"You know better than that, Danny. Tell me what you're really thinking."

"Yeah, you're right. For the last couple weeks, Sis and I have been focused on finding more cash in the business to keep things afloat. You helped us a ton. But it's tiring. She keeps pushing to talk about things like using our technology to help communities optimize their school bus routes, or helping adult learners get to the community colleges—stuff like that. It all seems out of reach to me. Right now, we need to figure out how to make payroll."

"It's kind of painful at times," Julie said.

"Sure is. I didn't know what Sis had been dealing with all this time."

"We had tough times too. We all do. If it hadn't been for keeping our vision in sight, I doubt we would have made it to where we are today. You must have a vision, a purpose, a reason for being in business to start with. We've got ours!" She grinned from ear to ear.

"Seems like you are super organized, Julie. We are still just scrambling."

"We haven't always been this way," she responded. "When I first met Sal, we were in bad shape. One of our vendors had some quality issues, and we had a serious injury on a client's job site due to a mechanical failure. Our QA team had not caught the problem in the shop before the robot went to the field. Thankfully, everyone was okay. It made us think hard about why we were in business to start with."

"That's kind of where we are right now."

"So, what are you going to do, Danny?"

"Well, we have a plan scratched out on the whiteboard back at the office."

"Let's talk about your plan. Maybe I can help you think it through," Julie offered.

"Okay, well, to start with, we are taking the advice you gave a couple weeks ago working through all our assumptions for the revenue waterfall. We have asked Mike, on our sales and marketing team, to focus all his time on generating more leads, writing content, tracking search engine rankings, social media, all that stuff."

"Sounds good so far."

"Next, we asked our tech support lead, Michelle, to start being proactive rather than reactive and set up quarterly business reviews with our customers. She has a neat process she uses for our customers. We learned about the QBRs from Eric over at Stream View."

"I know Eric. Great guy. We met at one of Sal's workshops."

"Oh, cool. Eric mentioned those workshops too. Michelle is also adding in some AI into our tech support, so our customers get better answers and get them quicker."

"Good idea. We did that—it made a huge difference. Our customers loved it."

"Okay. The last one is a doozy. Our R&D lead, Tommy T., is wrapping up development of some new AI tech that will allow us to get away from using lasers and just use image recognition. It's complicated stuff. I'm not sure it is worth all the money it will cost us. Once we get it done, we can start using that tech to do some of the stuff Sis wants to do, like the bus route optimization. She is passionate about figuring out how to help the cities we work with get more people access to education."

"That is starting to sound a lot like a true purpose to be in business, Danny. Is that something your customers want?" Julie asked.

"Yeah, I think so. For the last few years, Sis has been in sales and talking with prospects and customers full time. I think that is where she is hearing it. And from our dad. That was his idea too."

"Danny, I have an idea for you. I call it the vision test. It came in super handy back when we were struggling, and we still use it today. We start by creating a vision pyramid. On the bottom are the people—you know, because our people—"

"People are the foundation. Yeah. I been hearing that a lot lately."

"Well, it's the truth, so we start there. Then the next layer up in the pyramid is business strategies. Some people put initiatives or projects in this layer. It doesn't really matter. People first, then the strategies those people are working on. Then, at the top, we put our vision or purpose. The idea in the vision test is anytime you are making decisions related to either the people or the strategies, you simply ask yourself how the decision advances your vision. They all need to line up. The people line up to the strategies, and those strategies line up to the vision. If it doesn't, then consider that in your decision-making."

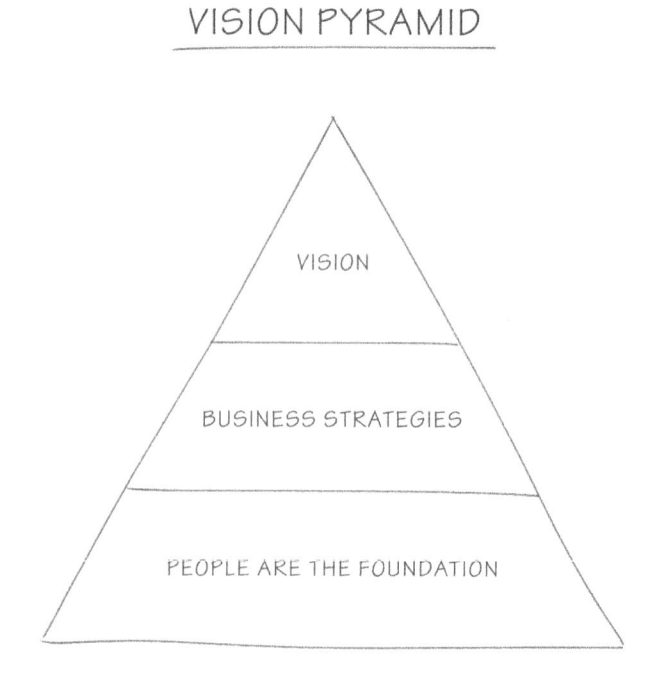

VISION PYRAMID

- VISION
- BUSINESS STRATEGIES
- PEOPLE ARE THE FOUNDATION

"Come on, it can't be that simple!"

"No, it's not, but it's a start. Think it through, Danny. Let's say you want to add Michelle's AI to tech support. Try it. Will this help us achieve our vision?" Julie said.

"Okay. If we add AI to tech support, we free up head count to work on other things, we get customers' questions answered faster, and the customers get better answers too," Danny said, with a little bit of lip service to the idea.

"Perfect—that's all good, right? Now, does that help you achieve Cecily's vision of communities increasing access to education?" Julie asked.

"Sort of, I guess, in a roundabout way, maybe."

"Let me take a shot at it, Danny. By using AI, you are learning about the technologies you will need for achieving the vision. If you use the freed-up time of the tech support staff to focus on vision-related activities, that will also advance the vision. The increased profitability can be used on vision-related activities to pay the software team as they develop the new products. Also, and maybe this is the biggest, you keep your customers happy, so they are still with you as you roll out the new products."

"Okay, I see your point, but can't you just do what you did there to justify anything?"

"No, not really. It doesn't always work. We were thinking about seriously reducing staff once, and when we did the vision test, we realized without enough staff, we were going to slow our pace down to a crawl and then maybe never achieve our vision. So, we looked at other avenues," Julie said.

"Got it! We were thinking the exact same thing. I had a big plan to reduce our R&D staff. I was all ready to present it to Sal, and before I could, he suggested we spend more on R&D."

"Yeah, that's Sal! He wants us all to be thinking about growth. All the time. Always going forward, never going back."

"Julie, I was planning on asking if you might be interested in buying some of our old inventory of lasers that we don't use anymore. But after our talk, I need to think a little more about how that decision fits with the vision thing."

"Good idea, Danny. Have you thought about donating them to someone like Mountainside School for their buses? You never know, that might be a better way to help achieve your vision."

"We must get you and Sis together to talk vision. Oh, by the way, she said she will be calling you about that leadership conference. I think she should definitely take you up on the offer."

"Awesome! Thanks for your help today too. Good to hear about your progress. I like your approach. And Cecily's vision. You've got this!"

"Thanks, Julie. That means a ton. See you soon."

Being at the school again made Danny start thinking about Ted. Was this Ted's challenge? Maybe he didn't have a vision or a purpose for his business.

Speak of the devil. Danny's phone rang. It was Ted.

***

JULIE
MOUNTAINSIDE SCHOOL

MEETING NOTES

VISION PYRAMID

→ PEOPLE ARE THE FOUNDATION

→ BUSINESS STRATEGIES
   - "INITIATIVES"
   - "PROJECTS"
   - "ACTIVITIES"

→ VISION / PURPOSE AT THE TOP

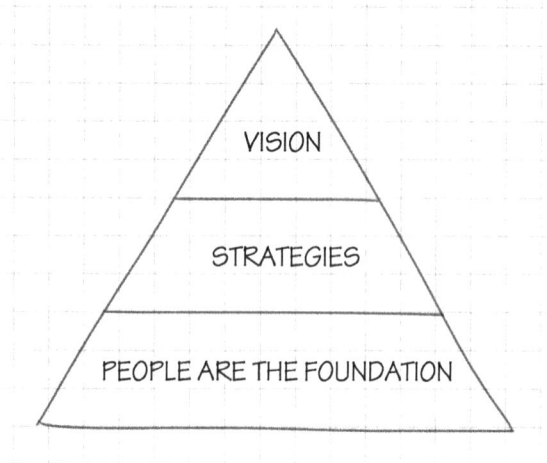

```
        /\
       /  \
      /VISION\
     /--------\
    / STRATEGIES\
   /-------------\
  /PEOPLE ARE THE \
 /   FOUNDATION    \
/_____\
```

THE "VISION TEST":

☆ "DOES THIS HELP US ACHIEVE OUR VISION?"

IF IT DOESN'T PASS THE TEST, DON'T DO IT!!!

"Did your date stand you up again?" the waitress asked as Danny waited for Ted to arrive at Johnny's Diner.

"Hey, Danny!" Ted said.

"Hey, bro, guess where I was yesterday. Mountainside School."

"No way. Haven't been back there since we were kids," Ted said.

"It's changed a lot. The chairs are a lot smaller."

"I want to hear all about it, but before you start, I've been thinking about our lunch last week."

"Yeah, I wondered why you wanted to get back together so soon."

"Here's the deal. My sales manager that quit last week—"

"Jessie?"

"Yeah, Jessie. Seems she didn't have another job lined up, after all. She just quit."

"That's odd. How do you know?"

"Her online profile says, 'open to work.'"

"I thought you said she quit to help some other company."

"I did, but maybe that wasn't it."

No doubt that was not it. It was probably Ted's leadership style, just like Danny was thinking.

"Ted, I always thought you said Jessie was sharp."

"Yeah, I know, she is, but she left me hanging. That's what I get for hiring one of Mr. Larobie's referrals."

"What? Mr. Larobie recommended Jessie to you?"

"Yep. See what I've been saying about him now? He said I could learn a lot from her or something like that."

"Sorry, Ted. Heck, maybe we need to hire her. We need sales."

"Sure, whatever, go ahead if you want to. She will probably leave you hanging too."

"Seriously, we might be interested."

"Yeah, of course. She knows sales, for sure. But I won't take her back now. She's on the market. Go for it."

"Okay, so anyway, you said you've been thinking?"

"Right, so I have to figure out why she really quit. Like I said the other day, she could have made more money working for me. Do you think she knows something I don't? Maybe about our competition?

Danny knew he should be gentle here. The meeting yesterday with Julie on vision and purpose could be a perfect fit.

"Ted, before we dig into that, I've been thinking about why Dad started Red Line back in the beginning. Cecily thinks he had a greater vision about helping communities grow."

"Calling her Cecily now, Danny?"

"Yeah, she leads the company now. The team respects her. I respect her. You should too. She has vision. That's where I was going with this. She has a vision for us."

"Leads the company, huh?"

"Well, we haven't talked about it yet, but she knows more about it than anyone else—the people follow her leadership. It is cool to watch. I'm still sorting out how I can help. Tell me a little bit about why you got into your business, Ted. What is your vision?"

"Okay, I'll play your vision game, Danny. I bought this company to make money. So, I could buy another company, and another, and get rich and drive a Lambo. Like why all people get into business. How's that for a vision? Oh, and throw a big house with a swimming pool in there too."

"I know you better than that, Ted. You don't want a Lambo. Seriously, can you remember why you chose this business to buy over any others you looked at?"

"Yeah. Let's see. The business broker I worked with convinced me it was undervalued. We could grow it and sell it for more. It was SBA financeable, which I needed, so that was great. I don't know. It was fintech software. I knew fintech companies were worth a lot. That's about it, I guess."

"Those sound okay, Ted. They aren't inspiring, though. Did anything about the business inspire you?"

"No, not really. Now that I think about it, the business is boring. Even when we are getting lots of sales, it is kind of boring. If it wasn't for the money, I probably wouldn't do it at all."

"What would you rather be doing? I know you love sports. Any way to use your business to work in sports?" Danny asked.

"Well, I thought once about building a budgeting app for student athletes who got injured and lost their scholarship money. That's just dreaming, though. Danny, you're grasping at straws. I want to figure out the real reason Jessie left. It doesn't have anything to do with whether or not I like the business."

"Really? It doesn't?"

"No! It must be something else. But I can't quite put my finger on it."

"Yeah, you're right. I am grasping at straws. The people I have been meeting with know how to connect to something bigger. Like a meaningful reason to be in business. Their people all seem so engaged."

"I get what you're saying, Danny. Like, I know our numbers okay enough. Our cash, our P&L—"

"You mean income statement."

"What?"

"Never mind. Forget I said that. Go on."

"Yeah, I know the numbers okay enough, but Mr. Larobie doesn't think I do. He asks me some simple questions about customer counts and average revenue per customer. All that info is right there in front of him in our quarterly reports. It's silly."

"Maybe that's not what he's really asking. Do you ask him any questions back? Like maybe what he would do, or what he thinks you should do?"

"No. Not really. I mean, why?"

"Ted, it's not just Mr. Larobie. There are lots of people out there who can and will help you. They're helping Sis and me. Sometimes all you need to do is ask. For instance, when Mr. Larobie asks about average revenue per customer, you could respond with the obvious answer and then follow up with a question about what he thinks a good average would be."

"Won't that just make me look like I don't know?"

"Well, do you know the answer?"

"No, but I don't want him to know that."

"Why not, Ted? He can't help you if he doesn't know what you need help with."

"I hadn't looked at it like that. Hey, how are you learning all this stuff so fast?"

"Mainly listening to smart people. And asking dumb questions."

"That should come easy for you, Danny!"

"Thanks, Teddie."

"Hey, don't call me that!"

"Something I plan to try is to put every idea through a vision filter. Simply take the idea and ask—does it help us achieve our vision?"

"Let me tell you my filter, Danny. Does it make me money!"

"Honestly, Ted, that's not bad. I like it. We are in a cash crunch right now, so that makes perfect sense for us too."

"Yeah, you're getting the idea. That's how I run the business. Every day. Does it make me money. It if doesn't, then I don't do it."

"Seems a little shortsighted, but I think we should add it into our planning."

"Planning? Never seem to get the time for planning."

"Oh no! Talk about timing. I almost forgot. I've got to get back to the office. Cecily has the team getting together. Ted, you helped me a lot today. Thanks, man! Gotta go!"

"Tell your boss, uh, Cecily, hi for me, Danny."

"Very funny, Teddie."

CHAPTER 9

The Missing Process

The whiteboard was blank in the Red Line conference room.

Michelle, Mike, Tommy T., and Sis were all waiting for Danny to return.

"Sis, what's going on? What happened to the plan?" Danny asked.

"We were waiting for you to get back so you could walk us through it," Sis said.

"Me? What about your vision and all the strategies we had listed?"

Michelle jumped in. "Yeah, that's exactly what we're talkin' about. We're waitin' for you to redraw it for us. You do a good job connectin' the dots, so we don't all get lost."

"Well, okay. Plus, I got a couple more ideas after talking to my friend Ted today."

"Ted? Really, Danny? Little Teddie?" Sis asked.

"Yeah, believe it or not, Ted has good ideas too!"

Danny grabbed a marker and got started. "Okay, let's start with vision. We'll use a triangle and put *vision* at the top. I got that from Julie over at Winning Robotics. So, Sis, give us the vision in as few words as possible. I'll write, you talk."

"It's simple. Our typical customer has a bunch of networked locations with either trains or buses connecting them. I envision us helping those communities to thrive by being more efficient with those networks. It has to do with helping underprivileged kids get a good education, helping people get access to healthy food, healthcare, social activities . . ."

"That's a lot of words, Sis! I don't have much space up here. How about something like 'We help communities offer efficient access to education, nutrition, healthcare, and social activities.' I'll put this over here to the right of the triangle. We can work out the wordsmithing later."

"The next rung down on the pyramid represents the strategies. We already have a few strategies we are working on. We might have to add a couple more over time. Okay, Mike, let's start with you. Whatcha got for us?"

"The paragraph summary I wrote up tells the story. Writing website content on the new technology for SEO. Hitting several social media channels to grow interest and engagement. And running some A/B tests on paid media. Not much, though, as there isn't any budget for it. Oh, we also are starting an email campaign to past leads in our CRM."

"Wow, that's a lot, Mike! Okay, cool, who's next?" Danny said.

"Wait a second. What's on the bottom rung on the pyramid, Danny?" Sis asked.

"That's the people part. Everyone I talk to says people are the foundation, so I put people at the bottom to support everything else. Who owns the strategy, maybe who they need to help them, stuff like that." Danny looked out over the group. "Okay, Michelle. You're up."

Michelle started. "First thing, we are rollin' out the chat tool to get customers better answers and faster. Then we are startin' to meet with our customers to talk about what they need and want from us. Actually, we are already doing both."

"So, the chat is this month?"

"Yep, almost done."

"How about the customer reviews?"

"We are scheduling them, but it is going to take a while to get through the whole list. Can we say ongoing?"

Tommy T. interjected, "Wait. So how are we going to measure this one? Having customer meetings doesn't mean it helps."

"Good point! That reminds me of something Ted suggested today. He has a test: 'How does this make us money?'–or maybe in our case, 'How does this get us cash?'"

"Yeah, that is exactly what I meant," added Tommy. "How does meeting with these customers get us any cash? How do we know if the meetings are working?"

"Yes, those are good questions, Danny," Sis agreed. "We'll add a customer success measurement part a little later. Now, how about you, Tommy?"

"Like Michelle said, we have the chat tool rolled out partially and we are tuning it against our knowledge base. It's working–still needs more tuning though. We need to add some tracking codes for Mike's A/B testing project, so put me down as a dependency on his work. We are fixing some dashboard bugs in the data analytics tool; that should make some customers happy. The main project is rolling out image recognition to replace the lasers. And it is a big one!" Tommy T. said.

"What do we do with the Red Line name if we don't have lasers?" Mike asked.

Tommy cut back in. "Oh, that's easy, the name didn't even come from the lasers."

"It didn't?" they all said almost in unison.

"Not at all. People have come to think that, but it originally came from the Red Line elevated train up in Metropolis City. That was our first customer. We had the technology but didn't even have a name for it. Once your dad got them to sign, he, your uncle Bill, and I decided to call the company Red Line Technologies. At that time, we didn't even think beyond one customer. Heck, we certainly never thought about image recognition and AI."

"That is all news to us, Tommy," Sis said as everyone laughed. The levity felt good for a change.

Mike, with a huge smile on his face, said, "This is too cool. We can drop the bright red line in our website and replace it with some kind of map, or city streets, or something that better fits with the vision."

Tommy T. jumped in. "Hey, Sis, thinking of maps, that reminds me. The other day, you asked if we could add locations of grocery stores to our data analytics maps. It was simple to add, so we tested it with one of our local customers' train and bus routes. Later we can show you the proximity of the stores to the stops. It was enlightening to see the gaps. Some people would have quite the walk to get their groceries."

"Perfect! Now that fits the vision for sure!" Sis said. "Danny, how about you take it from here. Write this up so it is simple and easy for us all to track. And it must connect the dots to the forecast. That is the only way this will all work."

Then she said, "Think you can get us a meeting with Sal and Phillip to get their feedback? I want us to run with this ASAP, but I want us to get it right too!"

"Sure, I can try," Danny said.

"Okay, let's break for the day and get back to business in the morning. Thanks for all the hard work today. It is going to pay off. Let's all make sure to hold each other accountable."

With that, Sis wrapped up.

"Sis, I'm not ready to go back to Sal and Phillip yet." Danny said after everyone else had left to head home.

"Why not, what's wrong?"

"I don't know exactly. Something's not right. I have Mike's marketing funnel metrics built like Eric showed us, but I just don't understand the close rates to get them from marketing into closed sales. I can draw the arrows to the forecast, but I can't nail the numbers. I just can't get the math in the forecast to work. I need a couple more days."

"Maybe there is someone in Sal's book we can call?"

"Maybe. Let me think about it. Not exactly sure what to look up, though. We can talk about that tomorrow. I need a little more time. Let me go down to my office and I'll think this plan through and get it drawn up."

"Your office?"

"Yeah, I'm hanging out with Tommy T.'s team."

"There's no office down there."

"I know, it's just one of the empty desks, but I like it. Close to the action. They are doing some cool stuff down there."

"Well, okay, but you can always stay up here in the conference room."

"Nah, you need this for your team meetings. I'll stay out of your hair this way. Maybe I can help Tommy T.'s team somehow with all the new stuff they're building."

"Okay, if you think so. See ya later, alligator," Sis said.

Sis was feeling more positive than Danny had remembered since he had been home from college. It was coming together. Almost.

***

Back at his desk in the R&D lab, Danny realized his forecasting spreadsheet had become more and more complicated. Tying his marketing numbers to his sales numbers was tough. There were a lot of assumptions to fit into the forecast. Danny hadn't wanted to say it to Sis since she led sales, but it was the sales part that didn't work. Danny had figured out Red Line didn't have much of a sales process. Not like Eric's marketing funnel or his uncle Bill's customer journey. Not even like Michelle's customer success process. It was another gap Phillip would surely point out.

Danny's uncle Bill had said a sales process needs lots of training and discipline. Danny didn't want to insult Sis, but he didn't see much discipline in her process. For as organized as she was, it didn't come through in sales, he thought.

THE MISSING PROCESS

Danny tried *By the Numbers*.

He looked up the different sales categories: *sales, sales compensation, sales presentation, sales tax, sales training*. No *sales process*. He decided to go with *sales training*.

"I have a question about the numbers—will you help me?"

Text sent.

Danny's phone rang almost as fast as he clicked send.

"Hi, how can I help you?" an energetic female voice said.

"Hi, wow, that was fast! My name's Danny and I'm with Red Line Technologies. I have a question about sales training, but, really, I think it's our sales process that needs help."

"Hi, Danny. I love these calls from Sal's book. I always learn something new. Happy to help you any way I can. I have worked with several of Sal's companies with their sales processes over the years. I just looked up Red Line. Are you out on River Road? I could swing by your office."

"Uh, sure! That would be great."

"How about eight o'clock tomorrow morning?"

"Sure, I'll be here."

"Okay, see you tomorrow."

This was happening way too fast for Danny. He needed to tell Sis. He knew she wanted him to use the book, but probably not for sales. That's her department. He hurried back up to Sis's office.

"Sis, I took your advice and used Sal's book."

"Awesome. That was fast. How did it work?"

"I guess okay. A lady from Sal's book will be here in the morning at eight."

"Why just okay? Everyone in Sal's book has been great."

"Well, the only listing in the book that made sense for the numbers I'm trying to work out was, um, for, uh, sales training. She said she has helped Sal's companies with their sales process too."

"I see. Well, we need that. To be honest, Danny, since Dad's been gone, I haven't closed a single new sale. We need a fresh set of eyes."

"Really? I thought you would be mad."

"No, silly, I'm the one that said go try the book. Now, who is she?"

"Uh, well, I don't know. I forgot to ask her name."

"Oh, brother," Sis said as she finished packing her bag for the night and headed toward the door.

***

Danny and Sis were sitting in the conference room at eight o'clock the next morning as the bell in the lobby rang.

Danny walked out to let in their guest.

"Hi, Danny, I'm—"

"Jessie?"

"Yeah, I'm Jessie. Good to meet you, Danny. Have we met before? You look familiar."

"We have. Well, just once. I'm a friend of Ted's. We met up at his office. Here, let's go into our conference room. My sister, Cecily, is waiting for us."

Danny had a million thoughts. Would Ted be mad? Was Sis okay with someone else looking over her shoulder? How did Jessie get in Sal's book?

"Hi, Cecily, I'm Jessie. Good to meet you!" Jessie said with a trademarked firm handshake and smile. "I was telling Danny on the phone how much I love getting these texts from Sal's book. I always learn something new."

"Thanks for meeting with us on such short notice. Danny tells me you have worked with several of Sal's companies," Sis said in her professional tone.

"That's right. I've been working with Danny's friend Ted as sales manager in his fintech business. Sal introduced me to Ted about a year ago."

"Jessie, before we get started, I should let you know, Ted told me you recently left his business."

"Yes, I did. I've learned a lot from the businesses and business leaders I work with, including from Ted. He has a strong, founder-led sales approach, which works well for his business. Where I can add the most value is by helping to implement a more formalized sales process when a company is ready to scale."

"We are definitely ready to scale. At least we think we are, or . . . we want to be ready to scale," Sis said.

"Mind if I ask you a few questions?" Jessie said.

"Sure, fire away." Danny jumped in, although somehow, he knew he wouldn't have much to add to this meeting today.

"Okay, tell me a little about your current sales process. Who leads the team, what type of leads do you get, what are some of the stages or phases a prospect goes through in the process?"

"We don't really have a process yet. Maybe that's our problem," Sis said.

"Well, Cecily, it might not be a problem at all. What I see from the companies I have worked with is they all actually do have a process, sometimes a very good process—it is just not documented. Some are four stages. Some are seven. It varies from company to company. The two biggest issues I see, though, are a lack of training on the process and then, even when there is training, a lack of discipline in adhering to the process."

"Okay, I get it. I know we need to be more disciplined. So, where do we start?" Sis asked.

"Let's start with where the leads come from. Mind if I use your whiteboard?"

"Yeah, sure. Go for it!" Danny added, handing her the only marker he knew wasn't already dry.

"I'll just draw a few empty boxes up here, and we can fill them in as we go. So, Cecily, where do most of your leads come from today?"

"Most of them are either from the website, like from organic search, or SEO. Then, of course, we have trade show leads. Those take longer to close. Oh, and we get referrals. Those are awesome when we get them. They almost always close."

"Perfect. When a lead comes in, what do you do first?"

"Well, first, I guess I listen to what the problem is they are trying to solve. We need to figure out quickly if our solution is going to be a fit for them. I don't want to waste their time, or mine, if it isn't going to be a fit."

Jessie wrote what Sis said in the first box, almost verbatim. "Let's call that *qualifying*. That sounds like your first phase in the process. What comes next?"

"After I do some research, I try to get them to see a demo of our product. I usually do a quick slide presentation too."

"Online presentation, like on a video call?"

"Yes, exactly. I like to ask a lot of questions as I go to make sure our solution really does solve their problem."

"Okay, great. Asking questions is good. I'll call that step *validation*. Something else important in a process that a lot of companies forget is having clear decision points to tell them when a prospect is ready to go the next step. Sometimes a salesperson will try to push them too fast and then lose them along the way."

"Yeah, I know I'm guilty of that. When we are getting desperate for a new customer, I forget to ask enough questions."

"Cecily, how do you currently track the progress of a prospect through the process?"

"Uh, we don't, really. Well, we have them in our CRM, and there are dates and times that I look back at. I guess you could say it's tracked that way."

"Actually, that is perfect. We might want to tweak the CRM some, but that is a perfect way to track. Most of the CRMs I have worked with will allow us to set up reports to spit out average times between stages in the process. We can also assign probabilities to close."

"I hadn't thought about giving them probabilities."

"It would be a little early for that now, but eventually, when you tracked enough deals through the process, you can start to use both your marketing funnel and your sales process to forecast your new sales."

Danny jumped in. "That is exactly what we are trying to do. We are building a forecast, and I've been having a hard time connecting our marketing funnel to closed sales."

"So, you think we are on the right track with this talk?" Jessie asked.

"Yes, of course. Let's keep going. I like the way you're thinking," Sis offered.

"Okay, great. We have *qualify* first, then *validate*. What do you do next, once you have validated that your solution is a fit through the demonstration and presentation?"

"I think this is where we tend to mess up. I always immediately send them a proposal with our pricing and timing. Sometimes I never hear back. That's when I know I did something wrong. But I don't know what."

"Totally get where you're coming from. I tend to mess that part up too. Let's call it the *proposal* stage. Instead of sending the proposal, have you ever tried walking them through it on a video call, like you did the demo?"

"No, I haven't tried that. Do you think they will do that? I always figured they wanted time to review it."

"If you position it as the way you always present your proposals, and this gives them a chance to ask questions in real time, then you will be surprised how many will agree. This gives you one more opportunity to see and hear their reaction as you walk through your solution in detail. It also gives you a chance to show them why you're the best fit."

"I get it. I've been doing sales for a long time. No one ever explained it to me quite like this. It's like we are making sure we're a fit at every step, before we go to the next step."

SALES PROCESS

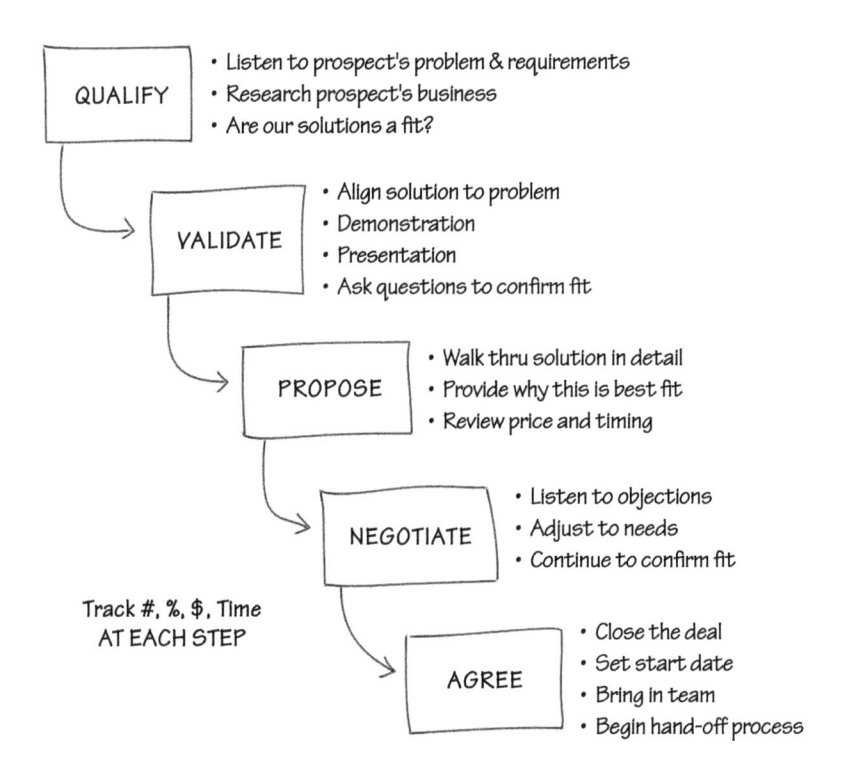

QUALIFY
- Listen to prospect's problem & requirements
- Research prospect's business
- Are our solutions a fit?

VALIDATE
- Align solution to problem
- Demonstration
- Presentation
- Ask questions to confirm fit

PROPOSE
- Walk thru solution in detail
- Provide why this is best fit
- Review price and timing

NEGOTIATE
- Listen to objections
- Adjust to needs
- Continue to confirm fit

Track #, %, $, Time
AT EACH STEP

AGREE
- Close the deal
- Set start date
- Bring in team
- Begin hand-off process

"That's exactly right, Cecily. Plus, you are making sure the prospect is agreeing you are a fit at each step too. If at any time you start to lose them, you will know. Almost before they do. Then you can adjust."

The meeting went on for several hours, as Sis and Jessie wrote up the rest of the process for negotiating and handing it off to Michelle in customer success. They took Jessie's ideas and added them to Sis's existing process.

It was fun for Danny to watch them go back and forth, teaching, then learning, going both ways. Listening, writing, erasing. Even starting over a couple times.

"Jessie, it has been amazing working with you today. I've learned a lot. Think we could do it again sometime? I'd like to write this up and get your feedback," Sis said.

"That would be great," Jessie replied. "Happy to help. Like I said when I got here, I always love getting these calls from Sal's book."

After Jessie left, Sis looked at Danny with a huge grin. He hadn't seen her this happy in a long time. Like a huge weight had just been lifted off her shoulders.

Danny was also feeling for his good friend Ted. He knew for sure now that Jessie was a winner, and Ted had messed up a good thing.

"Sis, let's talk. I think I now have what we need for the whole forecast. What you and Jessie just did was the missing process. Let me show you what I've got. It's pretty cool."

JESSIE + SIS
RED LINE CONFERENCE RM

MEETING NOTES

→ MOST COMPANIES HAVE A SALES PROCESS
    IT JUST MIGHT NOT BE DOCUMENTED YET

SALES PROCESS STAGES:

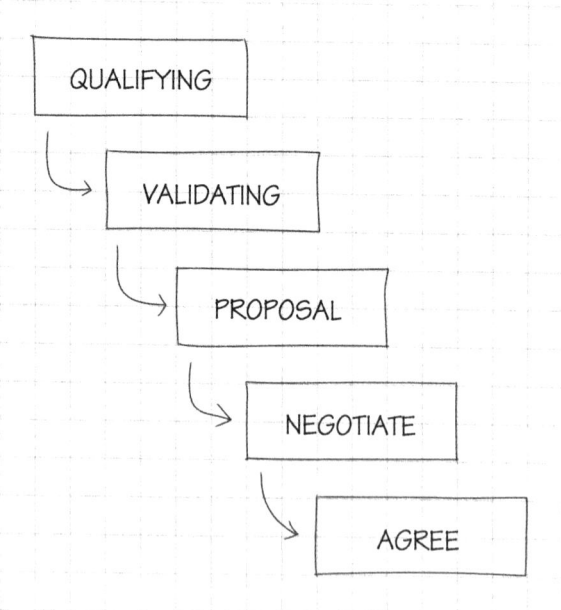

- QUALIFYING
  - → VALIDATING
    - → PROPOSAL
      - → NEGOTIATE
        - → AGREE

→ HAVE CLEAR DECISION POINTS TO GO TO NEXT STEP

☆ MAKE SURE YOUR SOLUTION IS A FIT AT EVERY STEP

TRAIN, TRAIN, TRAIN — DISCIPLINE!!!

CHAPTER 10

The Finishing Touch

"Let's start with the vision pyramid with our vision at the top, our strategies in the middle, and then our team as the foundation." Danny drew the pyramid to get the ball rolling.

"Okay that's good," Sis said. "Now let's add Uncle Bill's customer journey circle with all four of the process quadrants."

"Yep, this seems so simple now. First the R&D timeline section from Tommy T. Then the marketing funnel from Mike, followed by your new sales process. Finally, Michelle's customer success process." Danny was drawing fast as it all came together.

"Danny, now put the waterfall chart Julie showed us right next to the customer journey. That way, whenever we look at it, we can connect it back to the process and team member that owns each column. Danny, this is going to work!"

"I know. It's kind of scary, isn't it?"

"Yeah, I know. Danny, I'm so glad you came home to help me with all this. No way we are selling this business. We're going to grow it!"

VISION to FORECAST

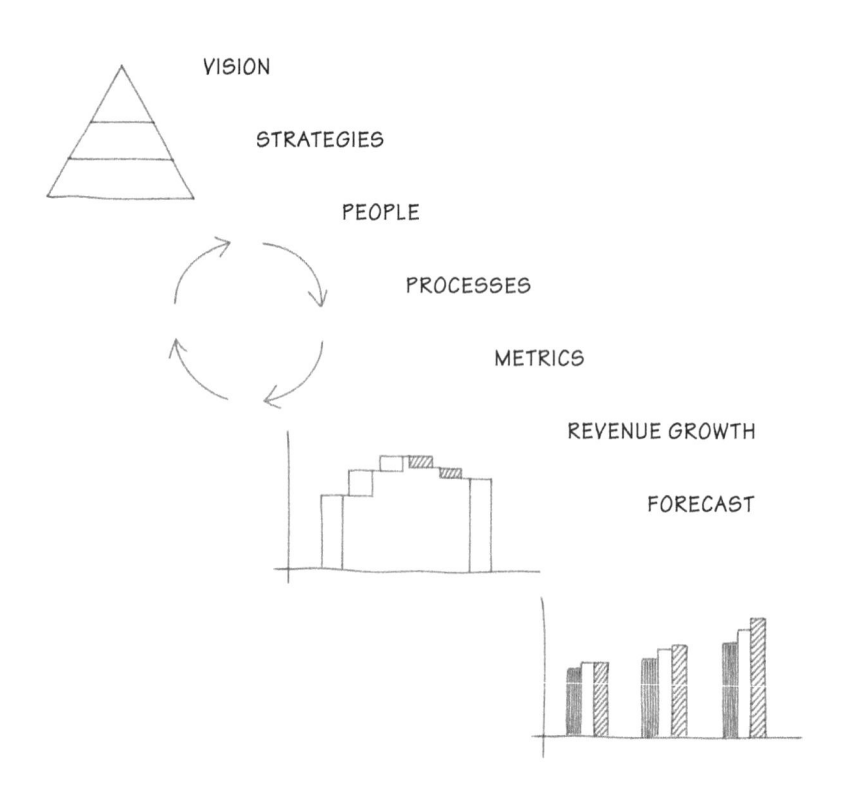

- VISION
- STRATEGIES
- PEOPLE
- PROCESSES
- METRICS
- REVENUE GROWTH
- FORECAST

"I'm in! Hey, Sis, I've got an idea. How about I show this all to Phillip before we present it? A couple people have told me he is good at shooting holes in forecasts."

"Sure, if you think you're up for it. See what he says, but let's get back in front of Sal as soon as we can. Let's take our team too. Let everybody see how united and excited we are."

***

Danny and Phillip had set 9:00 a.m. the next morning as the time. Danny was early, like he was taking a physics test at school. He couldn't wait to see if his and Sis's forecast would pass. Was the work good enough? Would his forecast math work? Would they be able to save the company?

"Danny, good to see you. Come on in!"

Going into Phillip's office was like an entirely different planet than being in Sal's old-school executive office with lots of wood and shelves loaded with books.

Phillip's office was bright and airy with lots of light. Like on the cover of *Architectural Digest*.

What caught Danny's eye were all the large flatscreens with spreadsheets and charts all over them. Like a stock trader in the movies.

There was a small working conference table for four people and a casual sitting area with low, modern leather chairs. This was cool! How did Ted not like this guy?

Danny and Phillip sat in the leather chairs and hung out for a while. Phillip asked Danny about what he liked most in college, what he liked least. How Danny had spent his free time, hobbies, golf, a bunch of stuff.

Danny learned Phillip had earned an undergrad mechanical engineering degree from MIT before he went to Stanford for his MBA. Phillip said he enjoyed engineering but decided he liked the business side of things better. Phillip was down-to-earth. For some reason, Danny had suspected all those MBA types would be pretentious jerks.

Phillip said he came back to Canyon Creek to be closer to family, but it sounded more like Sal offered him a deal he couldn't refuse.

"Looks like you printed up some new *By the Numbers* books." Danny noticed a stack on Phillip's desk that looked brand new.

"Yes. Actually, Danny, those are drafts of the updated version. We try to update them with new numbers a couple times a year. We aren't quite done with them yet. It might take a few more weeks, then we'll get you a new one to use. By the way, how is the book working for you?"

"It is awesome! Everyone we have called has been super helpful. They all seem to be open to hearing our ideas and asking us lots of questions. Not telling us what to do but listening to our questions too. Then talking it through. It's been great!"

"Good to hear. Like Sal says, to get in the book you must want to help others, and you also have to always be open and willing to learn more. In one way or another, everyone in the book started out just like you. Every story is a little different, just like every business is a little different. But there are some fundamentals we can all use."

"Probably not just tech either," Danny added.

"Right, the technology sector doesn't have any exclusivity on good business practices. Maybe just the opposite." Phillip laughed. "So, Danny, tell me about the journey so far. It's been a few days since we last met and we set up the shared drive."

"Your templates have been awesome, Phillip. We are trying to use the income statement you shared. Plus, the conversion calculations in the marketing funnel. It matched perfectly to what we learned from Eric over at Stream View Digital."

"Oh, you met Eric. That's great. He helps us with our workshops. Let's keep going. What else, Danny?"

"Can I show you the diagram of our strategies?"

"Of course, let's see it."

Danny jumped up to the glass whiteboard. It was much more professional than the one in the Red Line conference room that bounced on the wall when you wrote and looked like it had a hundred years of kids' coloring on it in the background. Probably Danny's when he would hang out waiting for his dad late at night.

The vision went up first with Julie's triangle, then the list of strategies with people as the foundation. So far, so good. Phillip was engaged. Danny added his uncle Bill's customer journey, Tommy T.'s R&D project timeline, then the marketing funnel he got from Eric, and the sales process Sis and Jessie made together. Then he wrapped up the circle with Michelle's customer success process. Finally, Danny showed Phillip the waterfall and the actual-to-forecast column chart.

"Danny, you made that all look easy. There is one thing I would add, though. We need to connect the waterfall to the real forecast in an income statement format. That way you

REVENUE FORECAST

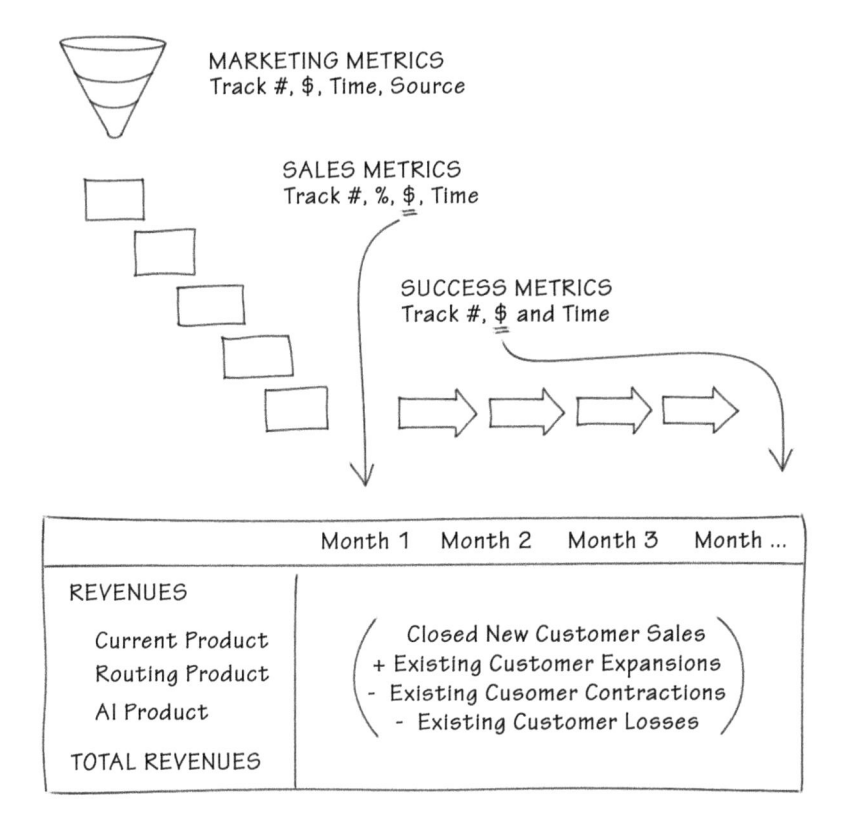

MARKETING METRICS
Track #, $, Time, Source

SALES METRICS
Track #, %, $, Time

SUCCESS METRICS
Track #, $ and Time

	Month 1	Month 2	Month 3	Month ...
REVENUES				
Current Product				
Routing Product	Closed New Customer Sales			
AI Product	+ Existing Customer Expansions			
	− Existing Cusomer Contractions			
	− Existing Customer Losses			
TOTAL REVENUES				

can make those actual-to-forecast charts come to life in your accounting software." Phillip added a blank income statement to the diagrams on the board.

"I mean this, Danny. I have never seen anyone dissect a business quite like you have, and in such a short amount of time. You would make a great analyst."

"Thanks. I guess with Cecily taking care of everything else, I was able to focus on it full time. I just tried to apply the same logic we had to use in my physics classes. No way we could have done this without the book, though!"

"Glad the book has helped. The connections are just a small part of it. You know that now," Phillip said.

Danny jumped in. "Oh, I almost forgot. Do you think we could set up another meeting with Sal to show him the forecast and what we've learned since our last meeting?"

"I was going to suggest the exact same thing. Would you bring Cecily and some of your team too?"

"That would be great. They might be a little nervous, though."

"Nothing to be nervous about. You guys seem passionate about your vision. It should be easy for them to talk about."

Phillip checked his calendar. "How about next Thursday? That will give you another week to polish the presentation."

CHAPTER 11

The Forecast

It had been exactly four weeks since Danny's very first meeting in Sal's office. The time had flown by. It seemed impossible for Danny and Sis to have made so much progress so quickly. At the same time, Danny knew they needed to be moving even faster. Some days, he didn't think they had made any progress at all. It had been a bit of a blur.

The gang was all there—Danny, Cecily, Michelle, Mike, and Tommy T. As well as Phillip and Sal, of course.

Phillip and Danny went back and forth on several drafts in the shared drive until they finally landed on the version for today.

Everyone had the printed version of the plan in front of them. Cecily had it up on the big screen too. It felt real now.

Not just scratches on a whiteboard. It was a plan. The Red Line plan.

"So, Danny, it's been a couple weeks since you first presented us your forecast. What have you learned and changed since then?" Sal asked.

"I can't believe it has only been two weeks! The entire thing has changed. Top to bottom. The main thing is the last plan was just revenues, without any good assumptions. Back then, we didn't know what we didn't know. This one is much more. It has a purpose. The financials are connected to a vision. One we all support." Danny looked around the room at the Red Line team.

"That sounds great, Danny. Let's hear all about it," Sal said.

"Cecily will walk us through it. She has been keeping us all on track the last few weeks. We couldn't have done this without her leadership," Danny admitted proudly.

Cecily started in. "Thanks, Danny. I'll start back with something our dad wanted more than anything else—helping communities grow and thrive. You see, he wanted our technology to do more than just count people coming and going from point A to point B. He wanted to help the community build efficient networks for people to thrive, to

THE FORECAST

help them get to and from educational facilities, to gain access to healthy food choices and economical healthcare. Our strategies in the plan will help us help them. With that, let's take a look at page . . ."

FORECASTING PROCESS

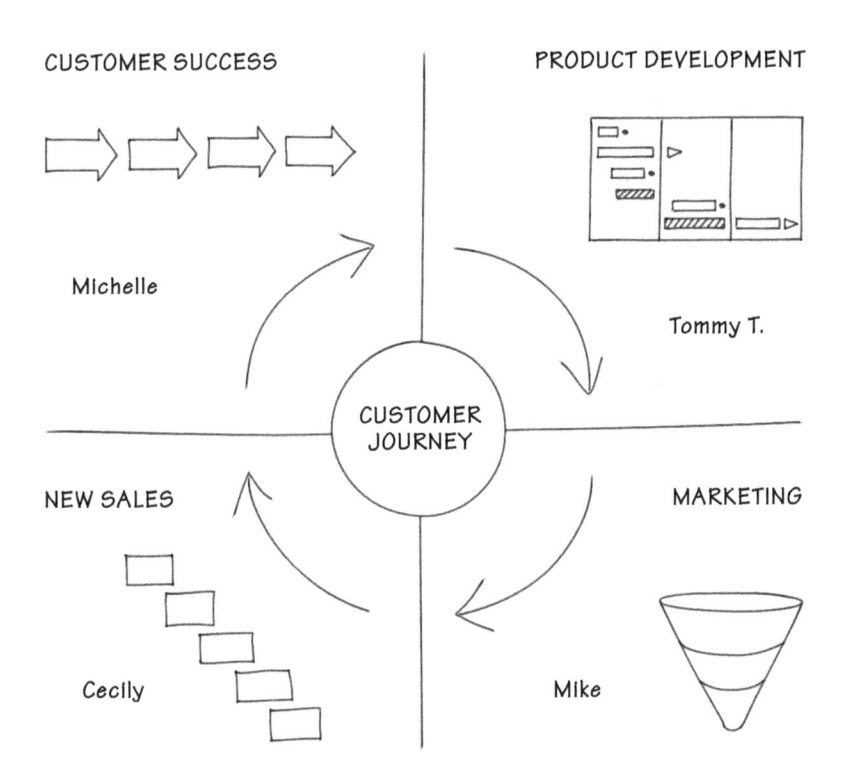

CUSTOMER SUCCESS

Michelle

PRODUCT DEVELOPMENT

Tommy T.

CUSTOMER JOURNEY

NEW SALES

Cecily

MARKETING

Mike

She nailed it! She went through the vision pyramid, laid out the key strategies. She had each person talk briefly about

STRATEGIES & METRICS

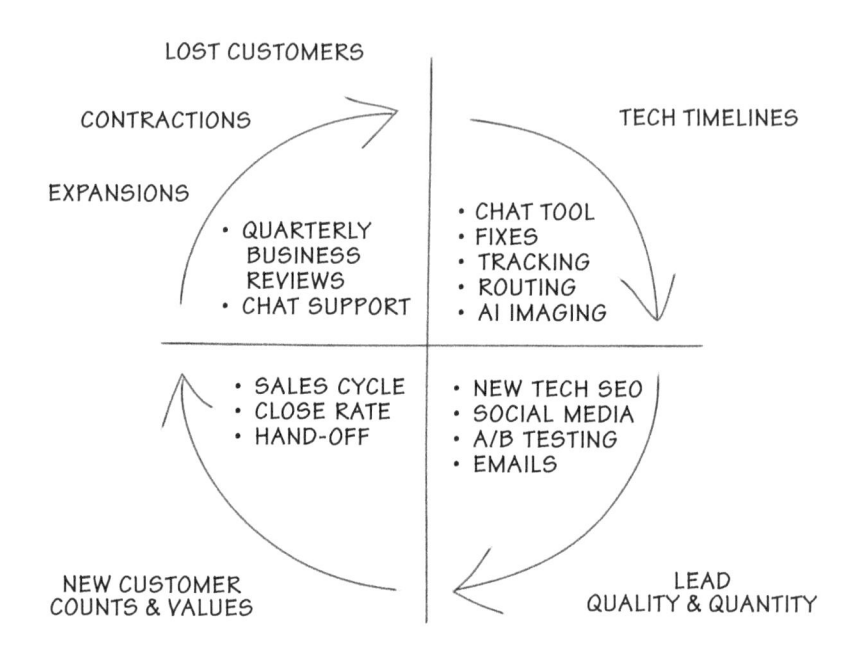

```
          LOST CUSTOMERS
                  |
  CONTRACTIONS    |        TECH TIMELINES
                  |
  EXPANSIONS      |
       • QUARTERLY    • CHAT TOOL
         BUSINESS     • FIXES
         REVIEWS      • TRACKING
       • CHAT SUPPORT • ROUTING
                      • AI IMAGING
  ─────────────────────────────────────
       • SALES CYCLE  • NEW TECH SEO
       • CLOSE RATE   • SOCIAL MEDIA
       • HAND-OFF     • A/B TESTING
                      • EMAILS
                  |
  NEW CUSTOMER    |        LEAD
  COUNTS & VALUES |        QUALITY & QUANTITY
```

their area, using Uncle Bill's customer journey circle. Phillip and Sal were asking questions, and everyone was engaged the whole way through. Danny thought it couldn't have been any better.

After the meeting ended, Sal and Cecily were talking off to the side. They both looked very happy, like the meeting went well. Danny wasn't sure specifically what they discussed, but Cecily was all smiles.

Phillip caught Danny on the way out. "Danny, great job pulling it all together. I can tell Sal was very impressed with your strategic thinking and Cecily's progress as a leader."

"Thanks, Phillip. All your templates and help made the difference."

"One more thing, Danny. Sal asked me to put you on his ninety-day calendar for an update session next quarter. Can we see you and Cecily again in three months? By the way, that is a very good sign, to be on Sal's ninety-day calendar."

"You bet! We would love that. Should be a fun three months."

Danny didn't know what he was saying.

***

Three months passed quickly.

Sal's boardroom was full this time, even more so than last. Cecily had invited Julie, Uncle Bill, and Eric too. They were official advisors to Red Line now. During the last three months, a lot had changed. It was hard to imagine just how much. Jessie joined the meeting too, as the interim Red Line sales manager. The room was packed.

THE FORECAST

Michelle had finally figured out the retention rates. Turns out they were worse than originally thought. Red Line was still losing customers to competing software solutions that didn't use hardware at all. Michelle said Red Line was late to market. The quarterly business review sessions had taught them a ton about what the customers wanted—and didn't want.

Cash was still extremely tight. New sales were not good either. However, Jessie was amazing. She had several customers ready to sign up for Tommy T.'s new image recognition to replace their lasers. Not until it was ready, of course.

On the flip side, Mike's marketing funnel worked great, with more leads than Danny ever thought possible.

Mike also landed Cecily some gigs on podcasts talking about how their technology could help communities with food deserts and access to healthcare.

And remember that customer who took legal action? Well, the other vendor they chose to implement instead of Red Line had failed to satisfy them as well. Cecily saved the day again with this one. She negotiated an early adopter rate for the new AI technology. They are testing now and apparently are huge fans of Tommy T. and his team. Even offered to do a case study on the new technology.

With all the new leads, and the new product about to roll out, the forecast for later in the year was looking good. Danny was both excited and exhausted at the same time. It had been a roller coaster ride.

Cecily started the meeting with all the learnings over the last three months: what the customers had shared in QBRs, how they had already started moving away from hardware, the market's need for their new technology, and how the leads were piling up. She closed the intro by saying that one of their customers, who was their strongest critic, was now first in line for their new tech. This seemed to excite the room, with lots of comments and smiles. Especially from Uncle Bill and Tommy T.

As Cecily went through the financials, Danny was disappointed the team didn't have better results to show. He just knew Sal and Phillip would be disappointed too. Danny thought it was depressing to see the team's hard work not getting them the growth. They had missed the forecast. Big time.

Cecily stayed incredibly positive through the whole presentation. Danny didn't know how she did it. She kept the team up all the time too. Everyone had their roles and knew how they were performing. There was alignment to the vision and strategies. She was good at this.

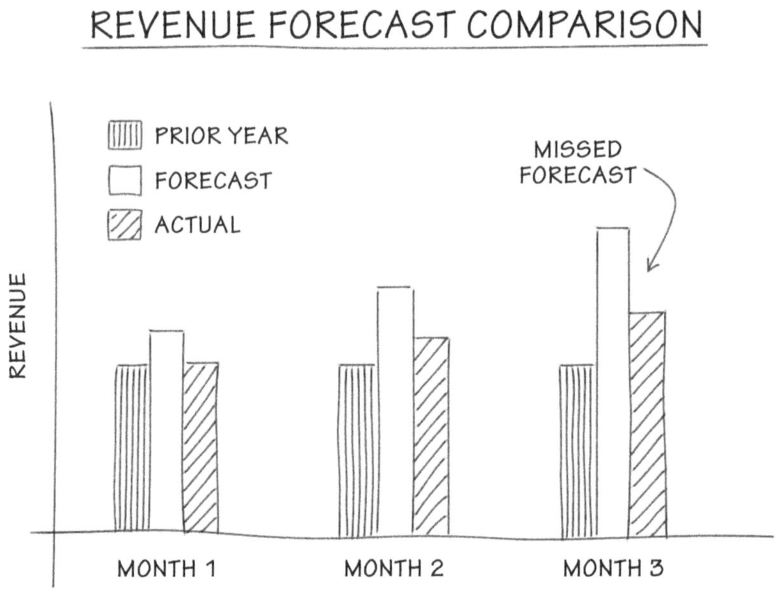
REVENUE FORECAST COMPARISON
PRIOR YEAR
FORECAST
ACTUAL
MISSED FORECAST
REVENUE
MONTH 1
MONTH 2
MONTH 3

As Danny sat there watching her, he wished he could help her more.

After the meeting, Sal came over to Danny. "Danny, I want you to know Phillip and I are very impressed with how you have pulled this all together and made the plan real."

"Thanks, but we missed our numbers, and not by just a little either."

"Don't sell yourself short. Just because the numbers aren't what you want doesn't mean you won't get them there. Keep your chin up. You're on a good team, Danny."

"Yeah, that's the best part. The team has really come together around the vision of where we are going."

"Also, I would like to ask you a favor. Would you present your forecasting process at our upcoming workshop for our portfolio companies? We'll call it 'More Than Numbers.'"

"You bet! I would love to." That pumped Danny back up.

After the meeting, as he sat in his car in the parking lot outside Sal and Phillip's office, it took a few minutes for it all to sink in.

Red Line had totally missed its numbers, but everyone was still positive and optimistic. He didn't get it. They missed the revenue targets. They missed the cash targets. Missed gross profit and net income too. Pretty weak quarterly report, he thought.

The forecast didn't work.

Or did it? Maybe there was truly more to it than just the numbers, like Sal said.

Red Line had a vision now and strategies to push it forward. There was a team that was clicking. The customers loved the new product direction.

About that time, Danny's phone buzzed. It was a text from Ted. Danny thought Ted probably wanted him to buy him lunch or coffee or something. It had been a couple of weeks.

Danny glanced down. The text said: "I have a question about how to use Sal's book *By the Numbers*. Will you help me?"

What?

No way!

Danny smiled as he imagined Phillip moving Ted's file from its old yellow folder over to a nice new blue one.

AFTERWORD

All forecasts are wrong. They are forecasts. It is something we must accept. What matters isn't whether the forecast is right or wrong. Of course, we all want to be as close to reality as possible with our forecasts. Accurate forecasting is a critical skillset needed in every leadership team. What matters most, though, is if the team is working together to learn and grow. Think of the forecast as a learning tool to help identify and capture opportunities for growth. Every variable and assumption become a lever to pull, a dial to adjust.

This story covered a lot of learnings beyond the technical aspects of forecasting revenues. The nuances of marketing funnels and sales processes are intertwined with the leadership lessons needed to build a long-term successful business. Those lessons are closely related, and I propose even required, to have a successful forecasting process.

Purpose for Being in Business

It cannot be understated, there must be a bigger purpose than making money. That is not to discount the importance of a profit motive; rather, it is to elevate profit to a higher level, a level with purpose. Over time, a business meanders. I prefer

AFTERWORD

to think of it as navigating. The business changes course, corrects for obstacles, seeks out less resistant paths, finds new customer segments, develops new products, and hires new employees. This navigation process also creates opportunities for expanding purpose. What drove the founder in the beginning startup phase may not be what is driving the business today.

As capitalization tables change from investments, private equity buyouts, mergers and acquisition transactions, and founders exit, it is common for a business to lose track of its purpose. It is a leader's role to build consensus around a shared purpose. Without purpose, the forecast is just numbers. We can all do better. Business is more than numbers.

Culture of Learning and Helping Others Learn Also

In a fictional fable like this one, it is all too easy to align with the advisors and mentors regarding a recent college graduate needing to learn business. What about the rest of us? Did our learning stop in our youth?

Learning is not a linear path from being the student to later in life becoming the teacher. Wisdom comes from experience. Some say from making mistakes. The wise among us know true wisdom comes from continuous learning. It never ends. Technology is always changing. The technology concepts in this book were out of date before the book was even printed. That is just the way it is.

While the financial frameworks contained in a three-statement financial model of the income statement, the balance sheet, and the statement of cashflow are much more stable, the business models of delivering software will change. The metrics of measuring business performance will change. The way we develop products and services, market, sell, and deliver customer satisfaction will change along with how rapidly the technology changes. This forces a business to learn and adapt new forecasting techniques.

Another one of the many roles of a leader is to build a culture of learning within the organization. It is learning, and helping others to learn also, that brings a forecast to life. Without this learning, the forecast will stagnate and die after just one period. We must learn and adapt to have growth.

Open to an Outside Perspective

Some have said true innovation comes from an outside perspective. I agree and will even expand that saying to include being open. True innovation comes from being open to an outside perspective. When we accept the forecast as a learning tool with levers and dials for tuning, then we must also accept that we have the responsibility, and even the obligation, to tune it. That requires being open to receiving feedback on what is working and what is not.

Feedback may come in the form of a marketing channel not producing as expected, or a sales process close rate weakening. It may be a straight-up comment directly from a customer during onboarding or activation. Feedback loops are everywhere. To have an effective forecasting process, we must be open and receptive to all these outside perspectives. This includes feedback from those pesky board members always pushing for growth.

Travel with the Customer

The story talks about the customer's journey and how to parallel the forecasting process to that journey: R&D building products to solve the customer's problem; marketing sharing content to align solutions with those problems; sales ensuring the alignment exists every step of the way through the sale process. Of course, customer success is making sure the customer is successful. Sounds simple, right? Well, sort of. At least it can be simple, in concept.

It is amazing the number of forecasts I've encountered over the years that were missed because of misalignment with customer expectations. The product built might not have truly solved the problem. Or, marketing attracted non-ideal prospects. Sales didn't listen, or worse, sales did listen and then discounted the feedback as incorrect or irrelevant.

Here's the deal. It is called a customer journey because it is a journey, not a destination once a contract is signed or a renewal is approved. We must be able to travel on that journey with the customer. Without this connection to the marketplace, the forecasting process is doomed to failure. We will miss our numbers.

Having a purpose, a culture for learning, and being open and receptive, all drive us to the same conclusion. It is the marketplace that decides our fate as business leaders. We cannot control how the market reacts, but we can control how we interact with the market. Listen and learn as you travel with the customer on their journey. Then pull those levers and twist those dials accordingly.

Final Thoughts

We don't know if Red Line Technologies will survive or not. That is not the point of the story. They are on the right path, though.

So many times, we as outsiders attempt to judge the success or failure of a business strictly on financial metrics. As the story exposes, it is not that simple. It is more than just the

AFTERWORD

numbers. Long-term success takes a vision, a purpose, a team, a culture, and a lot of discipline around the process.

What we do know is that Danny has learned a lot along his short forecasting journey. He can now take those learnings and apply them over and over again with whatever venture he chooses in the future.

Danny's best friend, Ted, has made the major transformation to being open and receptive. Of course, it will take much more time, but what a great first step.

Sis—excuse me, Cecily—is now in a role where she can apply her years of experience and leadership skills. Her desire and ability to help the team grow, and the customers be successful, sets her up for success.

Truth is, not all strategies will work out the way we planned. Some of the people will burn out. Some of the customers will switch vendors. Companies fail. It is part of life.

As Mr. Larobie—I mean Sal—and Phillip—aka "the Driver"—would remind us all, our job is to be open, learn from those experiences, and share our learnings with others. Most importantly, know that it is the people that make the numbers work.

There's a whole community out there to help.

APPENDIX

APPENDIX

SELF-ASSESSMENT QUESTIONNAIRES

As a good next step to improve your forecasting process, please consider completing the following self-assessment questionnaires. These questions align with the learnings in the story related to R&D timelines, marketing funnels, sales process, and the customer success process. Give it a try and set your company up for forecasting success.

FORECASTING PROCESS	167
RESEARCH & DEVELOPMENT FORECAST	169
MARKETING FUNNEL FORECAST	171
SALES FORECAST	173
CUSTOMER SUCCESS FORECAST	175

APPENDIX

FORECASTING PROCESS

Who on the forecasting team is responsible for pulling together the overall forecast?

Who are the team members involved in the forecasting process?

What tools are in place to develop and monitor actual results compared to the forecast?

What are the three to five key initiatives being implemented which will impact the forecast?

What are the specific timelines for delivering on these initiatives?

What is the biggest risk to effectively delivering on the forecast?

What is the biggest opportunity that can be captured in the upcoming period?

How does the leadership team receive feedback on the strengths or weaknesses in the forecast?

APPENDIX

RESEARCH & DEVELOPMENT FORECAST

Who on the forecasting team is responsible for tracking R&D timeline metrics?

What tracking tools are in place to track R&D timeline metrics?

What are the trigger points in the R&D timeline for telling marketing when to prepare?

What are the trigger points in the R&D timeline to tell sales when they can start selling?

What are the quantifiable ramifications to the forecast if the R&D milestones are missed?

How are data privacy and security concerns addressed to reduce risk to the R&D timeline?

How are defects in existing products addressed to reduce risk to the R&D timeline?

How does the R&D team receive customer feedback on the developed product?

APPENDIX

MARKETING FUNNEL FORECAST

Who on the forecasting team is responsible for tracking the marketing funnel metrics?

What tracking tools are in place to track marketing metrics?

What are the top three to five lead sources delivering the highest volume of leads in the forecast?

Which lead sources deliver the highest and lowest dollar value to customers?

Which lead sources deliver the highest and lowest percentage of won customers?

Which lead sources deliver the shortest and longest sales cycles?

What is the cost per lead associated with each lead source?

How does the marketing team receive customer feedback on alignment of marketing messaging with customer expectations?

APPENDIX

SALES FORECAST

Who on the forecasting team is responsible for tracking the sales process metrics?

What tracking tools are in place to track sales process metrics?

What are the phases in your sales process?

What are the specific criteria needed for a prospect to move from one phase to the next?

What is the overall average close rate?

What is the overall average sales cycle?

What are the seasonality assumptions, if any, in the close rate or sales cycle?

How are both won and lost prospects surveyed to learn and improve?

TO LEARN MORE

You can also access additional information on our website, including:

- Forecasting Readiness Self-Assessment
- Checklists for each of the four quadrants described in the customer journey
- Downloadable copies of the questionnaires
- Templates for conducting forecasting workshops

Please visit our website and contact us. We would love to hear your feedback.

www.RickRalston.com

ABOUT THE AUTHOR

Rick Ralston is an accomplished C-level executive and board director with a 25+ year technology career distinguished by the capacity to turn around and scale technology companies.

He is a down-to-earth pragmatic director, advisor, and mentor who bridges the gap between strategic vision and operational execution. Rick is respected in the boardroom as the credible voice of accountability.

His recent sector experience includes marketing tech, legal tech, and healthcare tech. Rick's past roles include CEO, President, COO, Global COO, Board Chairman, and Board Director. His experience is global and stretches across five continents.

Rick is well versed in vision creation, leadership development, investor and shareholder relations, and FP&A modeling. He has also led numerous M&A teams on both the buy-side and sell-side. Rick earned an MBA from The University of Nebraska at Lincoln and a BS in human resource management from Friends University in Wichita, KS.

Rick lives with his amazing wife, Robin, in the Kansas prairie east of Wichita.